MACROMOLECULAR CHEMISTRY

NEW RESEARCH

CHEMISTRY RESEARCH AND APPLICATIONS

CHEMISTRY RESEARCH AND APPLICATIONS

MACROMOLECULAR CHEMISTRY

NEW RESEARCH

VALENTIN GARTNER
EDITOR

New York

For permission to use material from this book please contact us:
Telephone 631-231-7269; Fax 631-231-8175
Web Site: http://www.novapublishers.com

NOTICE TO THE READER

The Publisher has taken reasonable care in the preparation of this book, but makes no expressed or implied warranty of any kind and assumes no responsibility for any errors or omissions. No liability is assumed for incidental or consequential damages in connection with or arising out of information contained in this book. The Publisher shall not be liable for any special, consequential, or exemplary damages resulting, in whole or in part, from the readers' use of, or reliance upon, this material. Any parts of this book based on government reports are so indicated and copyright is claimed for those parts to the extent applicable to compilations of such works.

Independent verification should be sought for any data, advice or recommendations contained in this book. In addition, no responsibility is assumed by the publisher for any injury and/or damage to persons or property arising from any methods, products, instructions, ideas or otherwise contained in this publication.

This publication is designed to provide accurate and authoritative information with regard to the subject matter covered herein. It is sold with the clear understanding that the Publisher is not engaged in rendering legal or any other professional services. If legal or any other expert assistance is required, the services of a competent person should be sought. FROM A DECLARATION OF PARTICIPANTS JOINTLY ADOPTED BY A COMMITTEE OF THE AMERICAN BAR ASSOCIATION AND A COMMITTEE OF PUBLISHERS.

Additional color graphics may be available in the e-book version of this book.

Library of Congress Cataloging-in-Publication Data

ISBN: 978-1-62417-854-2

Published by Nova Science Publishers, Inc. † New York

CONTENTS

PREFACE

In this book, the authors present current research in the study of macromolecular chemistry. Topics include efficient biotechological tools on GM crops and bioinsecticide approaches; structuring polymer composites with chemically active fillers; enrichment of Sauvignon Blanc wine by heat stable wine macromolecules; protein transduction in human cells mediated by arginine-rich cell-penetrating peptides in mixed covalent and noncovalent manners; and cleavage of fibrinopeptides from fibrinogen during fibrin formation.

Chapter 1 - The damage caused by insect–pests is one of the major causes of economic instability in the agribusiness. It is estimated that 20 to 40% of crop yield are lost annually due to insect attack, reaching more than 15 million dollars. Chemical pesticides are still not specific to combat insect-pests and are able to eliminate beneficial organisms to the agriculture. In addition, some chemical pesticides showed nocive side-effects to human beings and other animals, once that they pollute the environment irreversibly. The deleterious effects caused by the use of insecticides have incentivized the development of other alternatives, including the use of bacterial pathogens active against specific insects. Therefore, the bacterial species *Bacillus thuringiensis* (Bt), which present high specificity to insects, does not cause damages to the environment and their ability to keep the resistant levels controlled turned them into an agent for biological control agronomically important against several pests. The main reasons for the successful use of Bt include: (i) the diversity of proteins that are effective against a great variety of insect species; (ii) the security related to the non-target predators, parasites, birds and mammals; (iii) the facility of large-scale production at a low cost; (iv) the adaptability and the technology applied and (v) it is biodegradable. Nowadays,

Bts and Bt-derived products are highly represented on the market for insect control. During the last years, the introduction of genes into plants led to the development of the so called Bt crops and,in 2011, up to 42 million hectares was cultivated with GM crops (James, 2011). For Bt traits26% of the 160 million hectares of biotech crops grew in 29 countries, including Brazil. Therefore, this report intends not only to describe the features and uses of Bt toxins in agribusiness but also display recent innovations and techniques developed in Brazil, especially at the Research Institute EMBRAPA, in order to increase these proteins activities in the field, such as DNA Shuffling, Phage Display, Proteomic and Transcriptomic approaches.

Chapter 2 - At the present the main tasks in nanotechnology development are to combine the desired properties of the filler and polymer; to identify the factors which are determine the structure-properties relationship, as well as reinforcement effects of the filler embedded into the polymeric matrix. Nevertheless, a unique interaction between components within the composite structure complicates systematization. Despite of the considerable number of researches, the factors which determine strengthening of the final composite are unclear. Comprehensive study of the interface and dynamic effects during filler-filler and filler-polymer interaction will intensify understanding of the strengthening mechanism of the polymer based composites.

The modern theories of the polymer materials strengthening and factors which determine the structural effect of the silica fillers are discussed in this work. Influence of the filler surface modification on the polymer based composite structure is presented. Polymer matrices, filled with nanoparticles with mono and bifunctional grafted layer is in focus of this work.

It is show that the properties of the composite are conditioned by characteristics of the polymer layer near and on the surface of filler particles. A determinative at forming of composite structure is ability of the filler surface directionally to orient the growing in the process of polymerization macromolecules, and in case of co-polymers - to change also the composition of polymeric blocks due to selective adsorption of components of the monomers mixture. This work demonstrates that the final properties of the composite are conditioned by the structure of the polymer layer near and on the surface of filler particles. Ability of the filler to arrange monomers during polymerization is a key factor at the forming of composite structure; such arrangement is possible due to the selective adsorption of the initial component mixture.

Chapter 3 - Proteins and glycoproteins have shown to play a major role in heat induced haze formation in white wines despite of their low concentrations. The aim of this study was to improve thermal stability of white wine by enrichment with heat stable wine macromolecules. First, macromolecules from Sauvignon Blanc wine were separated and concentrated by sequential membrane fractionation, followed by the precipitation with acetone and then further purified by two-dimensional polyacrylamide gel electrophoresis. The main heat stable macromolecular fraction consisted of a glycosylated protein with molecular weight (MW) between 69 and 72 kDa and an isoelectric point of 3.25, which was identified by MALDI-TOF/TOF mass spectrometry as vacuolar grape invertase 1, GIN1. Using Concanavalin A affinity chromatography, high molecular weight compounds of 74 kDa and >250 kDa were retained, increasing the heat instability of wine from 48 NTU to 292 NTU. After desorption of these retained compounds followed by the enrichment of wine with glycoproteins having molecular weights of 70-100 kDa and >250 kDa, heat instability decreased by maximum 56%. However, this effect was mainly due to the desorption solution. Addition of grape invertase to Sauvignon Blanc would not improve wine stability.

Chapter 4 - Cell-penetrating peptides (CPPs) are small peptides with a high content of basic amino acid residues. They possess the ability to translocate through the plasma membrane and facilitate exogenous cargo delivery into living cells. In this chapter, we demonstrate that arginine-rich CPPs are able to not only traverse cellular membranes by themselves, but also carry macromolecules into human A549 lung carcinoma cells in mixed covalent and noncovalent manners. This special macromolecular delivery system was named as mixed covalent and noncovalent protein transductions (CNPT). We found that cells treated with nona-arginine (R9)-red fluorescent protein (RFP) fusion protein mixed with green fluorescent protein (GFP), referred to as R9-RFP/GFP complexes, exhibit both red and green fluorescent images. Cells treated with R9-GFP fusion protein mixed with RFP, denoted as R9-GFP/RFP complexes, emitted green and red fluorescence, vice versa. Furthermore, mechanistic studies revealed that the cellular uptake mechanism of CNPT may involve a combination of multiple internalization pathways. Therefore, applications of this binary CNPT system may provide an efficient tool for delivery of multiple proteins in bioscience and clinical research.

Chapter 5 - Fibrinogen, one of the most abundant proteins in the blood, plays a key role in haemostasis, inflammation, wound healing, and some other physiological and pathological processes.

In: Macromolecular Chemistry: New Research   ISBN: 978-1-62417-854-2
Editor: Valentin Gartner   © 2013 Nova Science Publishers, Inc.

Chapter 1

BT TOXINS: EFFICIENT BIOTECHNOLOGICAL TOOLS ON GM CROPS AND BIOINSECTICIDE APPROACHES

Patrícia Barbosa Pelegrini[1]
and Maria Fátima Grossi de Sa[1,2]
[1]Embrapa – Genetic Resources and Biotechnology,
Brasilia, – DF, Brazil
[2]Catholic University of Brasilia, Brasilia – DF, Brazil

ABSTRACT

The damage caused by insect–pests is one of the major causes of economic instability in the agribusiness. It is estimated that 20 to 40% of crop yield are lost annually due to insect attack, reaching more than 15 million dollars. Chemical pesticides are still not specific to combat insect-pests and are able to eliminate beneficial organisms to the agriculture. In addition, some chemical pesticides showed nocive side-effects to human beings and other animals, once that they pollute the environment irreversibly. The deleterious effects caused by the use of insecticides have incentivized the development of other alternatives, including the use of bacterial pathogens active against specific insects. Therefore, the bacterial species *Bacillus thuringiensis* (Bt), which present high specificity to insects, does not cause damages to the environment and their ability to keep the resistant levels controlled turned them into an agent for biological control agronomically important against several pests. The main reasons for the successful use of Bt include: (i) the diversity of proteins that are effective against a great variety of insect species; (ii) the security related to the non-target predators, parasites, birds and mammals; (iii) the facility of large-scale production at a low cost; (iv) the adaptability and the technology applied and (v) it is biodegradable. Nowadays, Bts and Bt-derived products are highly represented on the market for insect control. During the last years, the introduction of genes into plants led to the development of the so called Bt crops and,in 2011, up to 42 million hectares was cultivated with GM crops (James, 2011). For Bt traits26% of the 160 million hectares of biotech crops grew in 29 countries, including Brazil. Therefore, this report intends not only to describe the features and uses of Bt toxins in agribusiness but also display recent innovations and techniques developed in Brazil, especially at the Research Institute EMBRAPA, in order to increase these proteins activities in the field, such as DNA Shuffling, Phage Display, Proteomic and Transcriptomic approaches.

INTRODUCTION

Insect-pests present an important role in human life, acting as vectors for several diseases caused by parasites, as well as on the attack of plants and stocked products. Every year, there are many reports describing the loss in agriculture due to pest infestation. Crop damage due to insects, fungi and bacteria could account for up to 40% of total losses annually. The damage caused by insects is a major cause of economic instability in the agribusiness and large amount of the losses crop yield is due to insect-pests attack, reaching more than 15 million dollars.

The use of chemical insecticides has been the main practice for controlling insect-pests in agriculture worldwide. Nevertheless, the application of high quantity of pesticides in the field led to an increase of environmental pollution, including soil toxicity. Chemical pesticides are still not specific to combat insect-pests and some of them show noxious side-effects to human beings and other animals, due to pollute the environment irreversibly, as well as can lead to an intensification of human diseases, causing an enhance of cancer cases and disorders affecting the immune system (DEFRA, 2000; Pretty et al., 2000; Matlock and de la Cruz, 2002; Devine and Furlong, 2007).

Moreover, the use of chemical pesticides stimulated the improvement of some resistant insect populations, requiring the development of secondary strategies for insect control (Devine and Furlong, 2007). Hence, the use of microbial insecticides was offered as one of the replacements for chemical pesticides. However, it later showed to be limited in many aspects, including (i) narrowed range of activity; (ii) sensitivity to irradiation, variety of temperature, pH and oxygen income; (iii) generally, active only against the larval stage of the insects(Weinzierl et al., 2005; Devine and Furlong, 2007).

The deleterious effects caused by the use of insecticides have encouraged the development of alternatives, and the use of bacterial pathogens is the most widely known. Therefore, the bacterial species *Bacillus thuringiensis* (Bt) present high specificity towards insects, does not cause damages to the environment and their ability to keep the resistant levels controlled turned them into an agent for biological control agronomically important against several pests.

The main reasons for the successful use of Bt includes: (i) the diversity of proteins that are effective against a great variety of insect species; (ii) the security related to the non-target predators, parasites, birds and mammals; (iii) the facility of large-scale production at a low cost; (iv) the adaptability and the technology applied and (v) it is biodegradable. The strategy of using GM crops transformed with Bt toxin genes was so successful, that, from 2011, 90%of the land in U. S. designated to agriculture is now occupied with transgenic plants, including the ones containing proteins. Moreover, in Australia, 99.5%of all cotton cultivars are genetically modified, while in India, this crop reached 88%t (James, 2012). In addition, developing countries are increasing the area of biotech crops plantation in a proportion higher than industrial countries. Among them, Brazil is leading the land of planted area, behind only of U. S., with 30.3 million hectares being invested on Biotech crops, including maize, cotton and soybean. Compared to 2010, this extent increased 4.9 million hectares, making Brazil the leading country in the world on increasing genetically modified plants area (James, 2011). As it is going to be detailed further in this report, Brazil has approved 14 biotech crops during the years of 2010 and 2011, a total of 32 approvals since 2003, making it, again, the country with the fastest approval ratio for biotech crops worldwide (James, 2011). This is due, mainly, to its operative and advanced approval system for biotech crops, as well as to the partnerships between public and private organizations, including the Research Institute EMBRAPA, which contributes for the development of new biotech crops annually.

Hence, *Bacillus thuringiensis* (Bt) belong to a group of aerobic, Gram-positive and spore-forming bacteria, closely related to *Bacillus cereus* (Carlson et al., 1994). These bacteria were first isolated in Japan by Ishawata (1901) from *Bombyx mori* larvae and were considered an insecticide agent. Later, other lines of Bt were isolated in different places, including soil, dust, leave and stocked products (Burges and Hurst, 1977; Carozzi et al., 1991; Smith and Couche, 1991; Meadows et al., 1992).

Bacillus thuringiensis grows in a spore formation, when the nutrient amount is scarce in the medium. However, when the nutrient amount is abundant, the vegetative growth occurs normally. During sporulation, *B. thuringiensis* produces a high amount of insecticide proteins, which accumulate and form a stal that can reach 20-30% of the cell's total dry weight (Bechtel and Bulla, 1976). Although there are hundreds of Bt lines described nowadays, the number of different genes that each bacteria species carries widely varies. Furthermore, *B. thuringiensis* can produce several virulent components inside these stals, which allow them to growth inside susceptible or dead insect larvae. Among these molecules, there are not only proteins, but also phospholipases, proteases, chitinases, vegetative insecticide proteins (VIPs) and β-exotoxins (Levinson et al., 1990; Zhang et al., 1993; Estruch et al., 1997; Mac Innes and Bouwer, 2009; Agasthyaet al., 2012; Dong et al., 2012; Xiao et al., 2012; Gomaa, 2012).

The extraordinary diversity of toxins in nature seems to be related to the high level of genetic plasticity. The wide number of proteins permitted studies on sequence comparison and the identification of important elements for their function and specificity toward insect-pests (Hofte and Whiteley, 1989; Bravo et al., 1997; de Maggd et al., 2001; Kirouac et al., 2002; Masson et al., 2004; Puntheeranurak et al., 2004; Zhang et al., 2005; Vachon et al., 2012).

Hence, nowadays, more than 500 different gene sequences were described, in which the toxins were classified into 67 groups, according to their primary amino acid sequence, and the number is increasing day by day (Crickmore et al, 2010). This can be seen in a study in Spain, where more than 150 samples were analyzed, and several toxins from different groups active against diptera pest, *Ceratitis capitata,* were identified(Vidal-Quist et al., 2009). Members of this family consist of globular molecules containing three structural domains connected by simple interactions. To date, the tertiary structure of only seven proteins – 1Ac, 1Aa, 2Aa, 3Aa, 3Bb, 4Aa, 4Ba, 5Aa, 8Ea1, 1Ab19– were determined by x-ray crystallography (Grochulski et al., 1995; Derbshire et al, 2001; Galitsky et al, 2001; Boonserm et al, 2005, 2006; Guo et al. 2009; Li et al., 2001, 2009; Morse et al., 2001; Kashyap et al., 2012). These toxins demonstrated considerable differences in their amino acid sequences and specificity to insect species, and all of them show a high degree of similarity in the organization of the three domains, suggesting that the mechanism of action between them are the same (Bravo et al., 2004, 2011; Gómez et al., 2002, 2007; Zhang et al., 2005, 2006; Jurat-Fuentes and Adang,

2006; Pigott and Ellar, 2007; Pacheco et al., 2009; Likitvivatanavong et al., 2011; Vachon et al., 2012).

BT TOXINS: MECHANISM OF ACTION AND APPLICATIONS

Active toxins are able to bind to receptors at the surface of insect cells (BBMV), causing pore formation and death, and the characterization of this interaction has been extensively studied (Aronson and Shai, 2001; Bravo et al., 1992, 2004, 2011; de Maggd et al., 2003; Chattopadhyay et al., 2004;Jurat-Fuentes and Adang, 2006; Pigott and Ellar, 2007; Pacheco et al., 2009; Soberón et al., 2010; Whalon and Wingerd, 2003; Likitvivatanavong et al., 2011).

Therefore, the complex toxin/insect-receptor is the key factor for specificity and efficiency towards insect-pests. Hence, mutagenesis studies on three specific amino acid residues in the α-helix 5 of 1Ab (Lys157, Ser170 and Ser176) indicated that this entire structure was inserted into the midgut membrane of *Manduca sexta* cells, causing pore formation and insect death (Alzate et al., 2009). Similar results were observed for the control of mosquitos that are vectors for human diseases, such as *Aedes aegypti*. In this way, a protein receptor isolated at the midgut of the insect demonstrated synergistic features, increasing the toxic activity of 4Ba against the human vector (Park et al., 2009).

There are three main applications of Bt toxins: (i) control of insect-pests; (ii) control of mosquitos that are vectors for human diseases and (iii) development of transgenic plants resistant to insect-pests. One of the most successful applications of toxins is on the control of Lepidopteran, which are the most widely distributed class of insects around the world, including the United States, Canada, Brazil, India and China, among others. In all these countries, the control of insect-pests occurs basically by the use of Bt crops for the production of 1Aa, 1Ab, 1Ac and 2Aa toxins, present mainly in cotton and corn Bt crops(U. S. Sims et al., 1997; Environmental Protection Agency, 2001 Greenplate et al., 2001; Monsanto, 2003; Gouse et al., 2005, 2006, 2006a, 2009; Doyle et al., 2005; Kranthi et al., 2005; Clarck et al., 2005; Rochester et al., 2006; Gómez-Barbero, 2008; Barber, 2008).

Nevertheless, the development of transgenic cultivars expressing proteins has reached a higher stage on the replacement of chemical insecticides in the environment. The transformed Bt crops present a constant expression of proteins and maintain the toxins protected against degradation, facilitating their access to larvae and insects.

The production of toxins in plants has been improved through genetic engineering of genes, by removing sequences of splicing signaling, deleting the carboxi-terminal region of the protoxins and by utilizing the codon usage of the plant (Schuler et al., 1998; Bravo et al., 2007). The use of cultivars resistant to insects considerably reduces the acquirement of chemical pesticides in the field where transgenic plants are cultivated (Qaim and Zilberman, 2003). In addition, the use of transgenic cotton in countries such as China, Mexico and India has showed that these cultivars presented a significant positive effect on the yielding and application decrease of chemical pesticides (Kaphengst et al., 2011; James, 2010, 2011).

B. thuringiensis is involved in the production of large quantity of stal proteins, becoming, then, the essential host for development of *Cry* biopesticides. Some combinations of protein showed synergic effects on the control of insect-pests (Crickmore et al., 1995; Poncet et al., 1995; Lee et al., 1996). Therefore, it has been observed the use of different techniques of genetic manipulation on *B. thuringiensis* for the development of *cry* gene combinations and its application as bioinsecticides.

The advance on genetic engineering of *B. thuringiensis* started in the late 1980's, when diverse research groups developed and improved the technology to transform vegetative cell with plasmid DNA (Lereclus et al., 1989; Mahillon et al., 1989; Masson et al., 1989). The protocols differed in many steps, but each one could reach an average of 10^2 to 10^5 transformants per µg of plasmidial DNA. Later, the use of non-metilated DNA provided the obtaining of $3x10^6$ transformants (Macaluso and Mettus, 1991). Moreover, the variety of vector started being used on the introduction of *cry* genes cloned in *B. thuringiensis*. Alternatively, the integration of vectors for the insertion of *cry* genes by homologous recombination inside the resident plasmids or into chromosomes were also used. The application of this technique includes the rupture of *cry* and *cyt* genes in order to contribute for the pesticide activity (Poncet et al., 1993; Delecluse et al., 1993). The inactivation of protease production genes for the increase of stals formation and stability was also done (Poncet et al., 1993; Donovan et al., 1997).

Due to the many steps required for the processing and activation of the stal protein, it is not surprising that insect populations are now being able to develop ways to resist to the toxicity. It was earlier demonstrated that bioinsecticides based on Bt toxins are susceptible to promote resistance not only in laboratorial conditions but also in the field. However, the selection found inside laboratories can be very different from the resistance observed in the field, due to the lower level of genetic diversity encountered in the populations kept in laboratories. The mechanisms of resistance can be associated to natural selection, hence, any predator can be influenced by the development of Bt toxins resistance (Trisyono and Whalon, 1997). Reports evaluating insects/*Cry* toxins suggest that there are several mechanisms through which insects can reach resistance. The altered or inadequate processing of *Cry* toxins can be one of these mechanisms. Earlier studies described that the proteolitic activity in the midgut of *P. interpunctella* larvae resistant to Bt proteins was significantly reduced when compared to the midgut extract of susceptible insects (Oppert et al., 1994). Furthermore, a research with the Malasian insect *P. xylostella* - which is simultaneosly resistant to the subsepecies *B. thuringiensis kirstaki* and *aizawa*i – demonstrated that a unique locus coding a receptor common to several *Cry*1A toxins could be mutated, allowing the insect to be resistant to multiple toxins (Wright et al., 1997).

Studies involving the interaction of membrane receptors to Bt toxins labeled with I125 in BBMVs obtained from susceptible and resistant insects (*P. interpunctella* and *P. xylostella*) indicated that the binding sites to the toxin could be altered or modified in resistant lines (van Rie et al., 1990; Ferre et al., 1991). Nevertheless, in the case of resistant varieties of *Heliothis virescens*, neither the receptors of the binding affinity seemed to be affected (Gould et al., 1992). Therefore, these data suggest that the mechanism for resistance is complex and probably involves a post-binding event, such as the integration of the toxin to the inside part of the membrane or an activity in the ionic channels. The inadequate use of *Cry* toxins can rapidly increase the expansion of resistant insect species, wasting a precious source on the control of pests and agribusiness improvement.

IMPROVING RESEARCHES ON BT TOXINS

In order to develop new technologies for the control of several insect-pests, mainly for two of the most economically important cultures of Brazil – cotton and sugarcane – intensive research has been done at the Plant-Pest Laboratory (Embrapa, Brasília – Brazil). The primary focus was to use the germoplasm bank collection of *B. thuringiensis* strains from Embrapa Genetic Resources and Biotechnology, in order to identify novel toxins active against insect-pests.

In 2006, A Brazilian Normative Instruction (n$^{\underline{o}}$ 03/06) was issued in order to registrate microbial pesticides, demanding diverse toxicity assays to be carried out before registering the product. Among the tests required, there is the acute oral toxicity and pathogenicity analysis, which was previously standardized by the United States Environmental Protection Agency (USEPA 1996). The EMBRAPA found several new strains of Microbial pest control agents (MPCAs) to regulate insect-pests. In this way, the test of oral toxicity and pathogenicity was performed in five MPCAs, using mice in order to evaluate their potential as future biopesticides. Four strais were isolated from Brazilian soils, stored at a Colletcion of Entomopathogenic *Bacillus* spp. located at EMBRAPA Genetic Resources and Biotechnolocy, in Brazil and, later, tested against male and female C57BL/6 mice (Oliveira-Filho et al., 2009). The strains were: (i) two strains of *B. sphaericus* serotype H5, producing proteins of 51 and 42kDa, toxic to dipteran larvae (Monnerat et al., 2007); (ii) *B. thuringiensis* serotype *israelensisi,* expressing *Cry*4A, *Cry*4B, *Cry*11 and *Cyt*1 proteins, which are also toxic to dipteran larvae (Monnerat et al., 2005); (iii) *B. thuringiensis* serotype *kurstaki*, enconding *Cry*1Aa, *Cry*1Ab, *Cry*1Ac and *Cry*1B proteins, nocivous to lepidopteran larvae (Monnerat et al., 2007). After 30 days exposure, none of the bacteria strains showed toxicity or pathogenicity to the mice tested, indicating that these entomopathogens can be utilized in the environment as biopesticides (Oliveira-Filho et al., 2009).

Furthermore, from the strain S811 of *B. thuringiensis*, a gene denominate *cry1Ia12* was isolated, cloned into a vector and amplified (Grossi-de-Sa et alet al., 2007). The recombinant protein *Cry*1Ia12 was later expressed in *E. coli* and tested against the cotton boll weevil *Anthonomus grandis* and the fall armyworm *Spodotera frugipera.*While *Cry*1Ia12 was able to cause 50% mortality of *A. grandis* at a concentration of 230µg/ml, only 5µg/ml of the toxin was enough to cause the same rate of mortality when challenged against *S. frugiperda* larvae in *in vitro* assays (Grossi-de-Sa et alet al., 2007).

Later, *cry*1Ia12 gene was transformed into cotton plants in order to evaluate the insecticide properties of the Bt toxin *in vivo*. Qualitative *in vivo* assays using T1 generation plants revealed that transgenic cotton expressing *Cry*1Ia12 proteins had no effect on adults of *A. grandis*. However, after oviposition, no larvae survived or could be developed, demonstrating that *Cry*1Ia12 inhibit the growth of this insect larvae (data not published).

However, another concern of using *cry* proteins for the control of insect-pests is their specificity towards a certain targeted insect, not showing adverse effects on non-targeted animals, such as saprophytic insects and mammals. Numerous studies have demonstrated this particularity of *Cry* proteins on not affecting non-targeted animals (McClintock et alet al., 1995; Schrøder et alet al., 2007). Therefore, it is mandatory to analyze the potential risks of a new molecule when it is applied in a transgenic plant for the purpose of pest regulation. In this way, many organizations have released a safety assessment guideline to guarantee that food generated from biotechnology-derived products is as safe as those produced from conventionally breed crops. Among the Organizations, there is the Food Biotechnology Council (IFBC), the World Health Organization (WHO), the Food and Agriculture Organization of the United Nations (FAO) and the International Life Science Institute (ILSI). Hence, the effect of *Cry*1Ia12 was also estimated on growing rats to assess the risks of using this toxin on transgenic cotton plants (Guimarães et alet al., 2010). Thus, lab rats were fed with an artificial diet containing 100mg/Kg of *Cry*1Ia12 for ten days (average of 12mg/animal). Analysis of toxicology showed that *Cry*1Ia12 did not display adverse effects on rats. Furthermore, other evaluations confirmed these results, including relative dry weight of the animals internal organs, nutritional parameters, blood biochemistry profile and duodenum histological observations (Guimarães et alet al., 2010).

Knowing the potential of *Cry*1Ia12 as a novel tool against Leptopteran insects, innovative techniques were applied to this Bt toxin in order to discover if it could also be used against other insect-pests. Thus, the targeted-insect chosen was the sugarcane giant borer (*Telchin licus licus*), which is considered one of the most important cause of sugarcane loss in Brazil. Previous analyzes were done using *Cry*1Ia12 against *T. licus* without success, evidencing that this Bt protein was not toxic to the targeted insect. In this way, the production of variants with toxicity against *T. licus* was proposed, using DNA shuffling coupled with Phage-Display to generate a combinatorial library from the *cry*1Ia12 synthetic gene (Craveiro et alet al., 2010). Out of hundreds of sequences obtained, 30 were selected for toxicity assays, although only 4 of them exhibited toxicity towards the sugarcane giant borer. When the amino acid sequences were analyzed, eight local substitutions were observed in the active variants (Craveiro et alet al., 2010), suggesting that these regions might have relevant importance on the specificity and toxicity of *Cry* proteins towards insect-pests.

Moreover, a toxin gene from *B. thuringiensis* strain S1451 *cry*1Ia was isolated and the protein expressed and characterized for its insecticidal activity. *Cry*1Ia-type proteins had previously been described, showing ability to decrease insect larvae growth (Tailor et al., 1992), although theoretical tertiary structure presented a difference of three amino acid residues between both sequences, located at the loop region of the Domain III of the proteins

(Martins et al., 2008). *Cry*1Ia from *B. thuringiensis* S1451 also demonstrated similar structural features to *Cry*3A. Hence, the gene of Cry1Ia was introduced into baculovirus genome and infected on insect cells for protein expression. The recombinant *Cry*1Ia o was, then, evaluated towards *S. frugiperda* and *A. grandis* larvae, displaying toxicity abilities to both insect-pest larvae (Martins et al., 2008). The insecticidal activity against *A. grandis* of *Cry*1Ia from strain S1451 was 69 times higher than the activity demonstrated by *Cry*1Ia from strain S811 (Martins et al., 2007, 2008), suggesting the first *Cry*1Ia toxin can be selected as a good candidate for studies on transgenic plants resistant to the cotton bool weevil.

In addition, from the strain *B. thuringiensis* S811, another *Cry* toxin was isolated with activity against Coleoptera insects. The toxin denominated *Cry*8Ka1 showed ability to kill larvae of a population of *Anthonomus grandis* growth at Embrapa Genetic Research Laboratory (Magalhães et al., 2006). Then, in order to improve the activity of this Bt toxin against Coleopteran insects, the technique of DNA shuffling coupled with Phage-Display was performed using *Cry*8Ka1 as template (Oliveira et al., 2011). Among the *Cry*8Ka1 variants obtained, 200 hundred clones were selected, in which 30 clones were chosen based on expression level analyses (Oliveira et al., 2011). *Cry*8Ka1 and the elected variants were tested *in vitro* against neonate larvae of *A. grandis*. One of variants showed increased activity against *A. grandis*, when compared to the performance of *Cry*8Ka1 against the same insect-pest (Oliveira et al., 2011). While *Cry*8Ka1 could inhibit the growth of 50% of *A. grandis* larvae at a concentration of 8.93μg/mL, the variant protein was able to inhibit the same quantity of larvae at a concentration of only 2.83 μg/mL. For this reason, this variant was used for further assays and denominated *Cry*8Ka5 (Oliveira et al., 2011). Furthermore, the recombinant protein of this variant presented six major amino acid substitutions located in all three domains of the protein, as well as a deletion of 16 amino acid residues at its N-terminal. These alterations at the amino acid level provided alterations at the three-dimensional structure of the protein, which led to the increase in its activity towards Coleopteran insects.

Further studies were, then, performed in order to evaluate the specificity of *Cry*8Ka5 receptors located at the cell surface of insect midguts. Analysis of BBMV from *A. grandis* after interaction with *Cry*8Ka5 revealed the presence of a heat-shock cognate protein (HSP) and a vacuolar ATPase (V-ATPase) as receptor of this Bt toxin (Nakasu et al., 2010). Two-dimensional electrophoresis assay was also performed in order to confirm if these proteins (ATPase and HSP) play the role as surface receptor of *Cry*8Ka5. Five spots corresponding to protein-interacting with *Cry*8Ka5 were visualized, but only one sequence similarity to V-ATPase after MALDI-TOF-TOF analyzes (Nakasu et al., 2010). However, additional studies will be done on the identification of cell surface receptors for *Cry*8Ka5, once that N-aminopeptidases and cadherins – the classic *Cry* toxin receptors described in literature – were not detected in this report.

Similar investigation was carried out with *Cry* proteins from *B. thuringiensis* strain S601, called *Cry*1Ba6. This *cry* gene was earlier expressed and the purified protein showed high activity against the Coleopteran *A. grandis* at micromolar concentration (Monnerat et al., 2001). When a binding assay was performed using the midgut brush border membrane (BBMV) of *A. grandis,Cry*1Ba6 interacted with two proteins of 62 and 65kDa, respectively, homologous to alkaline phosphatases (ALPs) (Martins et al., 2010). This is the first time the localization of *Cry* proteins receptors are described in the midgut cells of *A. grandis*. Further experiments are being done now with the purpose of determining the protein sequence of these *Cry* toxin's receptors in *A. grandis*.

Besides those many *Cry* proteins analyzed, studies on the interaction of *Cry*1Ac with other molecules at the cell surface of insect midguts have also been carried out. Recently, the use of Bioinformatics tools coupled with a Lepidoptera Sequence Databank analysis permitted the identification of the gamma domain of a G-Protein (HvγGP) as a relevant membrane receptor for *Cry*1Ac (data not published). The three-dimensional structure of HvγGP was obtained and its docking with *Cry*1Ac was predicted through *in silico* assays. Moreover, Phage-Display technique was once more utilized to identify the binding sites of HvγGP and *Cry*1Ac. Results obtained here could indicate that the mechanism of action for *Cry* toxins leading to the activation of signaling pathways inside midgut cells might occurs through an interaction with HvγGP peptides, after oligomerization and insertion into the cell membrane (data not published). These results corroborate with earlier studies demonstrated by Jurat-Fuentes and Adang (2006), as well as by Zhang and cols (2006). Therefore, we suggest that *Cry* toxins, after a previous interaction with receptors located at the external surface of cell membrane, oligomerizes and inserts into the cell membrane, forming pores. The insertion permits an interaction of *Cry* toxins with G proteins located at the internal region of the cell membrane, releasing the α subunit of the complex, which will activate signaling pathways inside the cell that will lead to its death (data not published). Consequently, it will be possible to use the information of how *Cry*1Ac interacts with receptors at the cell surfaces to develop novel strategies and technologies to increase the performance of Bt toxins inside insect-pests organisms.

Simultaneously, another strain of *B. thuringiensis,* 1644, was used to isolate a *cry* toxin gene, denominated Cry1Ca. A truncated form of the Cry1Ca gene was transferred to baculovirus genome and the recombinant expressed protein was evaluated in cultured insect cells and on third-instar larvae of *S. frugiperda* (Aguiar et al., 2006). Using scanning electron microscopy, it was possible to observe the presence of cuboidal stal formations at the cells of *S. frugiperda* larvae infected with baculovirus containing Cry1Ca gene. It was previously suggested that stal formation by Bt toxins are due to the presence of Cysteine amino acid residues on the primary structure of the proteins (Bietlot et al., 1990). However, even with 8 cysteine residues less than the original Cry1Ca protein, the truncated protein was still able to form stal structures, indicating that the other residues might not be important for stal formation in this insect cells (Aguiar et al., 2006).

Later, the recombinant protein Cry1Ca was challenged against second instar larvae of *Anticarsia gemmantalis* and *S. frugiperda,* two important Lepidopteran pests of South America countries, such as Mexico and Brazil. While Cry1Ca was able to inhibit 50% of *S. frugiperda* larvae growth at a concentration of 114.4 ng/ml, only 19.49ng/ml of 1Ca was sufficient to decrease 50% of *A. gemmantalis* larvae growth (Aguiar et al., 2006). Although it has been reported that insect-pests are developing resistance towards proteins over the years (Ferre and van Rie, 2002), the results with Cry1Ca shows that the discovery of novel genes with insecticide activity can be an important strategy on developing gene banks and selecting candidates that can help on reducing the resistance of insects in field.

Also, the first report for the presence of a member of the 3-domain family of toxins in *B. sphaericus*, classically associated with *B. thuringiensis*, was also done with the participation of a research group from EMBRAPA, in Brazil. Hence, Cry48Aa1, the toxin from *B. sphaericus* related to stal toxins of *B. thuringiensis*, showed its mosquitocidal activity against *Culex quinquefasciatus* only when in the presence of a second toxin, Cry49Aa1 (from *B. thuringiensis* subsp. *israelensis* 4Q7) (Jones et al., 2007). The protein was, then, characterized as a toxin dependent of a second protein for activity toward mosquitos. It showed high identity with Cry4Aa – which can reduce the growth of *C. quinquefasciatus* larvae – although their mosquitocidal function showed different performances. The study also indicated that some genetic exchange may have occurred between bacteria from *Bacillus* gender, possibly using a mosquito host, giving rise to new toxin gene combination, displaying high relevance to pathogen-host coevolution (Jones et al., 2007). Further studies showed that the activity of Cry48Aa1 and Cry49Aa1 are specific to *C. quinquefasciatus*, once that it did not display any effects towards other mosquitoes, such as *Aedes* sp. and *Anopheles* sp., neither to Coleopteran, Lepidopteran or Dipteran insects (Jones et al., 2008).

Evaluations of non-targeted organisms are important to guarantee that toxins are specific to the insect pest of interest. Therefore, analysis of four Bt toxins – Cry1Aa, Cry1Ab, Cry1Ac and Cry2A – were performed against the zebrafish, *Danio rerio*, in order to confirm that these proteins are safe to use if released on water systems, such as lakes, rivers and ocean, as well as to investigate the toxicological effects of gene pyramiding in genetically modified Bt-crops (Grisolia et al., 2009). Thus, early stages and adults of zebrafish were challenged against all four proteins *in vitro*. The proteins were tested alone and on binary mixtures. Among the proteins analyzed, only Cry1Aa showed effect on zebrafish, enhancing its micronucleus frequency in peripheral erythrocytes of their adults. Moreover, none of the binary mixtures demonstrated activity towards zebrafish at a concentration of 100mg/L (Grisolia et al., 2009). However, all proteins displayed toxicity activity to embryos of zebrafish at a concentration varying from 25 to 150mg/L, retarding their development (Grisolia et al., 2009).

Furthermore, three other proteins from *B. thuringiensis* subsp. *Israelensis*, *Cyt*2Ba, Cry4Aa and Cry11A were tested against insect and mammalian cells *in vitro* in order to evaluate their toxicity level and possible synergic interaction (Corrêa et al., 2012). The cell lines used were: *S. frugiperda* IPLB-SF-21AE, *Lymantria dispar* IIPLB-LD-652Y, B*ombyx mori* BM-5, *Aedes albopictus* MCF-7 and human breast cancers cells MCF-7. Bt toxins were expressed individually and, then, tested separated and in binary combinations, before and after trypsin activation. Cry4Aa and Cry11A, both specific to dipteran insects, showed no toxicity towards Lepitopteran cells, but after trypsin cleavage, they demonstrated toxicity levels towards insect cells, although no effects were shown against mammal cells. Nevertheless, for *Cyt*2Ba, a toxicity activity was observed on human breast cancer cells (MCF-7), but no effect was seen towards insect cells. When evaluated as binary complex with Cry4Aa and Cry11A, the toxicity level displayed a slight increase, although not statistically significant (Corrêa et al., 2012). As the optimum pH for proteins activity is alkaline and the pH of human stomach is between 1 and 3, it is unlikely that these toxins will be functional when ingested by human beings.

CONCLUSION

Although very well studied and applied in many sources for the control of insect-pests, Bt toxins still present a wide range of alternatives to be discovered and developed on several pests whose regulatory mechanisms are not provided yet. Apart of bio-insecticides and Bt transgenic plants, side strategies can be used to optimize the regulation of several insect-pests, maintaining the specificity towards certain species without affecting other non-targeted organisms. That is why studies on Bt toxins continue to be relevant around the world, even when some pests are already showing resistance towards this protein group. Hence, the advance of new approaches - such as DNA shuffling and protein-protein interaction analysis - can provide valuable information about Bt toxins' mechanism of action, which can be used for the development of new molecules with enhanced activity against insect-pests. Therefore, there are still a lot to be discovered and applied on Bt toxins, as well as on their way of action against diverse organisms pathogenic to different plant cultivars. Bt toxin technology has been shown successful in diminishing the use of chemical insecticide and for the near future, novel genes for Bt toxin with improved insecticidal activities will be used alone or as gene stacking in order to control other important insect-pests in agriculture.

REFERENCES

Agasthya, A. S.; Sharma, N.; Mohan, A.; Mahal, P. Isolation and Molecular characterization of alkaline protease producing *Bacillus thuringiensis*. *Cell Biochemistry and Biophysics*, in print, 2012.

Aguiar, R. W. S.; Marints, M. E. S.; Valicente, F. H.; Carneiro, N. P.; Batista, A. C.; Melatti, V. M.; Monnerat, R. G.; Ribeiro, B. M. A recombinnant truncated 1Ca protein is toxic to lepidopteran insects and forms large cuboidal stal in insect cells. *Current Microbiology,* vol. 53, p. 287-292, 2006.

Alzate, O.; Hemann, C. F.; Osorio, C.; Hille, R.; Dean, D. H. Ser170 of *Bacillus thuringiensis* 1Ab -endotoxin becomes anchored in a hydrophobic moiety upon insertion of this protein into *M. sexta* brush border membranes. *BMC Biochemistry*, vol. 10, p. 25, 2009.

Aronson, A. I.; Shai, Y. Why *Bacillus thuringienis* insecticidal toxins are so effective: unique features of their mode of action. *FEMS Microbiology Letters,* vol. 195, p. 1-8, 2001.

Barber, J. Laval survival in Bollgard cotton. *The Australian cottongrower,* vol. 29, no. 2, 2008.

Bauce, E.; Carisey, N.; Dupont, A.; van Frankenhuyzen, K. *Bacillus thuringiensis* subsp *kustaki* aerial spray prescriptions for balsam fir stand protection against spruce budworm (Lepidoptera: Torticidae) *Journal of Economical Entomology,* vol. 97, p.1624–1634, 2004.

Bechtel, D. B.; Bulla, L. A. Jr. Electron microscopic study of sporulation and parasporal stal formation in Bacillus thuringiensis. *Journal of Bacteriology,* vol. 121, p. 1472–1481, 1976.

Becker, N. Bacterial control of vector-mosquitoes and black flies. In: Charles J. F., Delécluse A., Nielsen-le Roux C., editors. *Entomopathogenic bacteria: From laboratory to field application.* Kluwer Academic Publishers, p. 383, 2000.

Bietlot, H. P. L.; Vishnubhatla, I., Carey, P. R., Pozsgay, M.; Kaplan, H. Characterization of the cysteine residues and disulfide link ages in the protein stal of *Bacillus thuringiensis. Journal of Biochemistry,* vol. 267, p. 309-315, 1990.

Boonserm, P.; Davis, P.; Ellar, D. J.; Li, J. stal structure of the mosquito-larvicidal toxin 4Ba and its biological implications. *Journal of Molecular Biology,* vol. 348, no. 2, p. 363-382, 2005.

Boonserm, P.; Mo, M.; Angusthanasombat, C.; Lescar, J. Structure of the functional form of the mosquito larvicidal 4Aa from *Bacillus thuringiensis* at a 2.8-angstrom resolution. *Journal of Bacteriology,* vol 188, no. 9, p. 3391-3401, 2006.

Bravo, A.; Hendrickx, K.; Jansens, S.; Peferoen, M. Immunocytochemical analysis of specific binding of Bacillus thuringiensisinsecticidal stal proteins to lepidopteran and coleopteran midgutmembranes. *Journal of Invertebrate Pathology,* vol. 60, p. 247-253, 1992.

Bravo, A. Phylogenetic relationships of *Bacillus thuringiensis* δ-endotoxin family proteins and their functional domains. *Journalof Bacteriology,* vol.179, p. 2793–2801, 1997.

Bravo, A.; Gómez, I.; Conde, J.; Muñoz-Garay, C.; Sánchez, J.; Miranda, R.; Zhuang, M.; Gill, S. S.; Soberón, M. Oligomerization triggers binding of a *Bacillus thuringiensis* 1Ab pore-forming toxin to aminopeptidase N receptor leading to insertion into membrane microdomains. *Biochimica et Biophysica Acta,* vol. 1667, p. 38-46, 2004.

Bravo, A.; Sarjeet, S. G.; Soberón, M. Mode of action of *Bacillus thuringiensis* and *Cyt* toxins and their potential for insect control. *Toxicon*, vol. 49, no. 4, p. 423-435, 2007.

Bravo, A.; Likitvivatanavong, S.; Gill, S. S.; Soberón, M. *Bacillus thuringiensis*: a story of a successful bioinsecticide. *Insect Biochemistry and Molecular Biology*, vol. 41, p. 423-431, 2011.

Burges, H. D.; Hurst, J. A. Ecology of *Bacillus thuringiensis* instorage moths. *Journal of Invertebrate Pathology*, vol. 30, p. 131-139, 1977.

Carlson, C. R.; Caugant, D. A.; Kolstø, A. B. Genotypic diversity among *Bacillus cereus* and *Bacillus thuringiensis* strains. *Applied Environmental Microbiology*, vol. 60, p. 1719-1725, 1994.

Carozzi, N. B.; Kramaer, V. C.; Warreen, G. W.; Evola, S.; Kozid, M. G. Prediction of Insecticidal activity of Bacillus thuringiensis strains bypolymerase chain reaction product profile. *Applied Environment Microbiology*, vol. 57, p. 3057-3061, 1991.

Chattopadhyay, A.; Bhatnagar, N. B.; Bhatnagar, R. bacterial insecticidal toxins. *Critical Review in Microbiology*, vol. 30, p. 33-54, 2004.

Clarck, B. W.; Phillips, T. A.; Coats, J. R. Environmental Fate and Effects of *Bacillus thuringiensis* (Bt), Proteins from Transgenic Crops: a Review. *Journal of Agriculture and Food Chemistry*, vol. 53, p. 4643-4653, 2005.

Corrêa, R. F. T.; Ardisson-Araújo, D. M. P.; Monnerat, R. G.; Ribeiro B. M. Cytotoxicity analysis of three *Bacillus thuringiensis* subsp. *Israelensis* d-endotoxins towards insect and mammalian cells. *PLOS One*, vol. 7, no. 9, p. 1-9, 2012.

Craveiro, K. C.; Gomes Júnior, J. E.; Silva, M. C. M.; Macedo, L. L. P.; Lucena, W. A.; Silva, M. S.; Souza Júnior, J. D. A.; Oliveira, G. R.; Magalhães, M. T. Q.; Santiago, A. D.; Grossi-de-Sá, M. F. Variant 1Ia toxins generated by DNA shuffling are active against sugarcane giant borer. *Journal of Biotechnology,* vol. 145, p 215-221, 2010.

Crickmore, N., Zeigler, D. R., Schnepf, E., van Rie, J., Lereclus, D., Baum, J., Bravo, A., Dean, D. H., 2010. *Bacillus thuringiensis toxin nomenclature.* http://www.biols. susx.ac.uk/Home/Neil_Crickmore/Bt/index.html.

DEFRA (Department of Environment, Food and Rural Affairs). Department for Environment, Food and rural Affairs: *Design of a Taz or Charge Scheme for Pesticides*. Retrieved from http://www.defra.gov.uk/environment/chemicals/pesticides/pesticidestax on September 14, 2005.

Delécluse, A.; Poncet, S.; Klier, A.; Rapoport, G. Expression of IVA and IVB genes, independently or in combination, in a stal-negative strain of

Bacillus thuringiensis subsp. *israelensis. Applied Environmental Microbiology*, vol. 59, p. 3922-3927, 1993.

Derbyshire, D. J.; Ellar, D. J.; Li, J. stallization of the *Bacillus thuringiensis* toxin 1Ac and its complex with the receptor ligand *N*-acetyl-D-galactosamine. *Acta Crystallography*, vol. D57, p. 1938-1944, 2001.

Devine, G. J.; Furlong, M. J. Insecticide use: contexts and ecological consequences. *Agriculture and Human Values,* vol. 24, no. 3, p. 281-306, 2007.

Dong, F.; Zhang, S.; Shi, R.; Yi, S.; Xu, F.; Liu, Z. Ser-substituted mutations of Cus residues in *Bacillus thuringiensis* VIp3Aa7 exert a negative effect on its insecticidal activity. *Current Microbiology,* vol. 65, no. 5, p. 583-588, 2012.

Donovan, W. P.; Tan, Y.; Slaney, A. C. Cloning of the nprA gene for neutral protease A of *Bacillus thuringiensis* and effect of *in vivo* deletion of nprA on insecticidal stal protein. *Applied Environmental Microbiology*, vol. 63, p. 2311–2317, 1997.

Doyle, B.; Reeve, I.; Coleman, M. The Cotton Consultants Australia 2005 Bollgard Comparison Report. A Survey of Cotton Growers' and Consultants' Experience with Bollgard in the 2004-2005 Season. *A report prepared for the Cotton Research and Development Corporation and the Cotton CRC.* 136pp, 2005.

Estruch, J. J.; Carozzi, N. B.; Desai, N.; Duck, N. B.; Warren, G. W.; Koziel, M. Transgenic plants: an emerging approach to pest control. *Nature Biotechnology*, vol. 15, p. 137-141, 1997.

Ferre, J.; Real, M. D.; van Rie, J.; Jansens, S.; Peferoen, M. Resistance to the *Bacillus thuringiensis* bioinsecticide in a field population of *Plutella xylostella* is due to a change in a midgut membrane receptor. *Proceedings of the National Academy of Science USA.*, vol. 88, p. 5119-5123, 1991.

Ferre, J., van Rie, J. Biochemistry and genetics of insect resistance to *Bacillus thuringiensis. Annual Reviews in Entomology,* vol. 47, p. 501-533, 2002.

van Frankenhuyzen, K. Application of *Bacillus thuringiensis* in forestry. In: Charles J. F., Delécluse A., Nielsen-le Roux C, editors. Entomopathogenic bacteria: from laboratory to field application. *Kluwer Academic Publishers,* p. 371, 2000.

Galitsky, N.; Cody, V.; Wojtczak, A.; Ghosh, D.; Luft, J. R.; Pangborn, W., English, L. Structure of the insecticidal bacterial δ-endotoxin 3Bb1 of *Bacillus thuringiensis. Acta Crystalographica Section D: Biological Crystalography,* vol. 57, p.1099–1101, 2001.

Gould, F.; Martínez-Ramírez, A.; Anderson, A.; Ferré, J.; Silva, F. J.; Moar, W. J. Broad-spectrum resistance to *Bacillus thuringiensis* toxins in *Heliothis virescens*. *Proceedings of the National Academy of Science USA,* vol. 89, p. 7986-7990, 1992.

Gomaa, E. Z. Chitinase production by *Bacillus thuringiensis* and *Bacillus licheniformis*: their potential in antifungal biocontrol. *Journal of Microbiology,* vol. 50, no. 1, p. 103-111, 2012.

Gómez, I., Sáncehz, J.; Miranda, R.; Bravo, A.; Soberón, M. Cadherin-like receptor binding facilitates proteolytic cleavage of helix alpha-1 in domain I and oligomer pre-pore formation of *Bacillus thuringiensis* 1Ab toxin. *FEBS Letters,* vol. 513, p. 242-246, 2002.

Gómez, I.; Pardo-López, L.; Muñoz-Garay, C.; Fernández, L. E.; Pérez, C.; Sánchez, J.; Soberón, M.; Bravo, A. Role of receptor interaction in the mode of action of insecticidal and *Cyt* toxins produced by *Bacillus thuringiensis*. *Peptides*, vol. 28, p. 169-173, 2007.

Gómez-Barbero, M.; Berbel J.; Rodriguez-Cerezo, E. Adoption and performance of the first GM crop introduced in EU agriculture: Bt maize in Spain. *Joint Research Center*. Sevilla: European Commission, 2008.

Gouse, M.; Pray, C. E.; Kirsten, J.; Schimmelpfennig, D. A GM subsistence crop in Africa: the case fo Bt white maize in South Africa. I*nternational Journal of Biotechnology*, vol. 7, no. 1/2/3, p. 84-94, 2005.

Gouse, M.; Piesse, J.; Thirtle, C. Output and Labour Effects of GM Maize and Minimum Tillage in a Communal Area of Kwazulu-Natal. *Journal of Development Perspectives*, vol. 2, no. 2, p. 71-86, 2006.

Gouse, M.; Pray, C.; Schimmelpfennig, D.; Kirsten, J. Three seasons of subsistence insectresistant maize in South Africa: Have smallholders benefited? *AgBioForum*, vol. 9, no. 1, p. 15-22, 2006a.

Gouse, M.; Piesse, J.; Thirtle, C.; Poulton, C. Assessing the performance of GM maize amongst smallholders in KwaZulu-Natal, South Africa. *AgBioForum*, vol. 12, no. 1, p. 78-89, 2009.

Guo, S.; Ye, S.; Liu, Y.; Wei, L.; Xue, J.; Wu, H.; Song, F.; Zhang, J.; Wu, X.; Huang, D.; Rao, Z. stal structure of *Bacillus thuringiensis* 8Ea1: an insecticidal toxin toxic to underground pests, the larvae of *Holotrichia parallela*. *Journal of Structural Biology*, vol. 168, no. 2, p. 259-266, 2009.

Greenplate, J.; Mullins, W.; Penn, S.; Embry, K. 1Ac levels in candidate commercial Bollgard® varieties as influenced by environment, variety and plant age: 1999 geneequivalent field studies. *Proceedings of the Beltwide Cotton Conference*, vol. 2, p. 790-793, 2001.

Grisolia, C. K.; Oliveira, R.; Domingues, I.; Oliveira-Filho, E. C.; Monnerat, R. G.; Soares, A. M. V. M. Genotoxic evaluation of diferente □-endotoxins from *Bacillus thuringiensis* ion zebrafish adults and development in early life stages. *Mutation Research*, vol. 672, p. 119-123, 2009.

Grochulski, P.; Masson, L.; Borisova, S.; Pusztai-Carey, M.; Schwartz, J. L.; Brousseau, R.; Cygler, M. Bacillus thuringiensis IA(a) insecticidal toxin: stal structure and channel formation. *Journal of Molecular Biology*, vol. 254, p. 447-464, 1995.

Grossi-de-Sa, M. F.; Magalhães, M. Q.; Silva, M. S.; Silva, S. M. B.; Dias, S. C.; Nakasu, E. Y. T.; Brunetta, P. S. F.; Oliveira, G. R.; Oliveira Neto, O. B.; Oliveira, R. S.; Soares, L. H. B.; Ayub, M. A. Z.; Siqueira, H. A. A.; Figueira, E. L. Z. Susceptibility of *Anthonomus grandis* (Cotton boll weevil) and *Spodoptera frugiperda* (fall armyworm) to a 1Ia-type toxin from a brazilina *Bacillus thuringiensis* strain. *Journal of Biochemistry and Molecular Biology*, vol. 40, no. 5, p. 773-782, 2007.

Guillet, P.; Kurstack, D. C.; Philippon, B.; Meyer, R. In: Bacterial Control of Mosquitoes and Blackflies. de Barjac, H.; Sutherland, D. J. Editors. Rutgers Univ. Press; NJ: p. 187–190, 1990.

Guimarães, L. M., Campello, C. C.; Rocha, T. L.; Vasconcelos, I. M.; Carvalho, A. F. U.; Mulinari, F.; Grossi-de-Sá, M. F. Short-term evaluation in growing rats of diet containing *Bacillus thuringiensis* 1Ia12 entomotoxin: nutritional responses and some safety aspects, *Journal of Biomedicine and Biotechnology*, p 1-8, 2010.

Hofte, H.; Whiteley, H. R. Insecticidal stal proteins of *Bacillus thuringiensis*. Microbiol. Molecular Biology Review, vol. 53, p. 242-255, 1989.

Ishawata, S. On a kind of severe flacherie (sotto disease). *Dainihon Sanshi Kaiho,* vol. 9, p. 1-5, 1901.

James, C. Double-digit growth continues for biotech crops worldwide. Ithaca, NY: *International Service for the Acquisition of Agri-biotech Applica tions.* 2004.

James, C. Global Status of Commercialized Biotech/GM Crops: 2010. ISAAA Brief No. 42.Ithaca, NY: *International Service for the Acquisition of Agri-biotech Applications.* 2010.

James, C. Global Status of Commercialized Biotech/GM Crops: 2011. ISAAA Brief No. 43. Ithaca, NY: *International Service for the Acquisition of Agri-biotech Applications.* 2011.

James, C. USDA Crop Acreage Report for 2012, confirms that US farmers continue to demonstrate overwhelming trust and confidence in biotech crops. Manila: *International Service for the Acquisiton of Agri-biotech Applications*. 2012.

Jones, G. W.; Nielsen-Leroux, C.; Yang, Y.; Yuan, Z.; Dumas, V. F.; Monnerat, R. G.; Berry, C. A new toxin with a unique two-component dependency from *Bacillus sphaericus. The FASEB Journal,* vol. 21, p. 4112-4120, 2007.

Jones, G. W.; Wirth, M. C.; Monnerat, R. G.; Berry, C. The 48Aa-49Aa binary toxin from *Bacillus sphaericus* exhibits highly restricted target specificity. *Environmental Microbiology*, vol. 10, no. 9, p. 2418-2424, 2008.

Jurat-Fuentes, J. L.; Adang, M. J. toxin mode of action in susceptible and resistant *Heliothis virescens* larvae. *Journal ofInvertebrate Pathology*, vol. 92, p. 166-171, 2006.

Kashyao, S.; Singh, B. D.; Amla, D. V. Computational tridimensional protein modeling of 1Ab19 toxin from *Bacillus thuringiensis* BtX-2. *Journal of Microbiological Biotechnology,* vol. 22, no. 6, p. 788-792, 2012.

Kaphengst, T.; Nadja E. B.; Clive E.; Robert F.; Sophie H.; Stephen M.; Nataliya S. Assessment of the economic performance of GM crops worldwide. *Report to the European Commission,* March 2011.

Kirouac, M.; Vachon, V.; Noël, J.-F.; Girard, F.; Schwartz, J. L.; Laprade, R. Amino acid and divalent ion permeability of the pores formed by the Bacillus thuringiensis toxins 1Aa and 1Ac in insect midgut brush border membrane vesicles. *Biochimica et Biophysica Acta*, vol. 1561, p. 171–179, 2002.

Kranthi, K. R.; Naidu, S.; Dhawad, C. S.; Tatwawadi, A.; Mate, K.; Patil, E.; Bharose, A. A.; Behere, G. T.; Wadaskar, R. M.; Kranthi, S. Temporal and intra-plant variability of 1Ac expression in Bt-cotton and its influence on the survival of the cotton bollworm, Helicoverpa armigera (Hübner) (Noctuidae: Lepidoptera). *Current Science*, vol. 89, no. 2, 2005.

Lee, M. K.; You, T. H.; Young, B. A.; Cotrill, J. A.; Valaitis, A. P.; Dean, D. H. Aminopeptidase N purified from gypsy moth brush border membrane vesicles is a specific receptor for Bacillus thuringiensis IAc toxin. *Applied Environmental Microbiology*, vol. 62, p. 2845-2849, 1996.

Lereclus, D.; Arantes, O.; Chaufaux, J.; Lecadet, M.-M.Transformation and expression of a cloned d-endotoxin gene in *Bacillus thuringiensis. FEMS Microbiology Letters*, vol. 60, p. 211-218, 1989.

Levinson, B. L.; Kasyan, K. J.; Chiu, S. S.; Currier, T. C.; Gonzalez, J. M. Jr. Identification of b-exotoxin production, plasmids encoding b-exotoxin, and a new exotoxin in *Bacillus thuringiensis* by using high-performance liquid chromatography. *Journal of Bacteriology*, vol. 172, p. 3172- 3179, 1990.

Li, J.; Carroll, J.; Ellar, D. J. stal structure of insecticidal delta-endotoxin from *Bacillus thuringiensis* at 2.5A resolution. *Nature*, vol. 353, p. 815-821, 1991.

Likitvivatanavong, S.; Chen, J. W.; Evans, A. M.; Bravo, A.; Soberon, M.; Gill, S. S. Multiple receptors as targets of toxins in mosquitoes. *Journal of Agriculture and Food Chemistry*, vol. 59, p. 2829-2838, 2011.

de Maagd, R. A.; Bravo, A.; Crickmore, N. How *Bacillus thuringiensis* has evolved specific toxins to colonize the insect world. *Trends in Genetics*, vol. 17, no. 4, p. 193–199, 2001.

de Maagd, R. A.; Bravo, A.; Berry, C.; Crickmore, N.; Schnepf, H. E. Structure, diversity, and evolution of protein toxins from spore-forming entomopathogenic bacteria. *Annual Review in Genetics*, vol. 37, no. 409– 433, 2003.

Mac Innes, T. C.; Bouwer, G. An iproved bioassay for the detection of *Bacillus thuringiensis* beta-exotoxin. *Journal of Invertebrate Pathology*, vol. 101, no. 2, p. 137-139, 2009.

Macaluso, A.; Mettus, A.-M. Efficient transformation of*Bacillus thuringiensis* requires nonmethylated plasmid DNA. *Journal of Bacteriology*, vol. 173, p. 1353-1356, 1991.

Mahillon, J.; Chungjiatupornchai, W.; Decock, J. Transformation of Bacillus thuringiensis by electroporation. *FEMS Microbiology Letters* vol. 60, p. 205–210, 1989.

Magalhães, M. T. Q. *Toxinas Cry: Perspectivas para obtenção de algodão transgênico brasileiro.* Dissertation in Cellular and Molecular Biology, Universidade Federal do Rio Grande do Sul, Brazil. 2006.

Martins, E. S.; Praça, L. B.; Dumas, V. F.; Silva-Wemeck, J. O.; Sone, E. H.; Waga, I. C.; berry, C.; Monnerat, R. G. Characterization of *Baciluus thuringiensis*isolates toxic to cotton boll weevil (Anthonomus grandis). *Biological Control*, vol. 40,p. 65-68, 2007.

Martins, E. S.; Aguiar, R. W. S.; Melatti, V. M.; Falcão, R.; Gomes, A. C. M. M.; Ribeiro, B. M.; Monnerat, R. G. Recombinant 1Ia protein is highly toxic to cotton boll weevil (Anthonomus grandis Boheman) and fall armyworm (Spodoptera frugiperda). *Journal of Applied Microbiology*, vol. 104, p. 1363-1371, 2008.

Martins, E. S.; Monnerat, R. G.; Queiroz, P. R.; Dumas, V. F.; Braz, S. V.; Aguiar, W. S.; Gomes, A. C. M. M.; Sánchez, J.; Bravo, A.; Ribeiro, B. M. Midgut GPI-anchored proteins with alcaline phosphatase activity from the cotton boll weevil (Anthonomus grandis) are putative receptors for the 1B protein of *Bacillus thuringiensis*. *Insect Biochemistry and Molecular Biology,* vol. 40, p. 138-145, 2010.

Masson, L.; Prefontaine, G.; Peloquin, L.; Lau, P. C. K.; Brousseau, R. Comparative analysis of the individual protoxin components in P1 stals of *Bacillus thuringiensis* subsp. *kurstaki* isolates NRD-12 and HD-1. *Biochemistry Journal*, vol. 269, p. 507–512, 1989.

Masson, L.; Schwab, G.; Mazza, A.; Brousseau, R.; Potvin, L.; Schwartz, J. L. A novel Bacillus thuringiensis (PS149B1) containing a 34Ab1/35Ab1 binary toxin specific for the western corn rootworm Diabrotica virgifera virgifera LeConte forms ion channels in lipid membranes. *Biochemistry*, vol. 43, p. 12349–12357, 2004.

Matlock, R. B.; de la Cruz, R. Na inventory of parasitic Hymenoptera in banana plantations under two pesticide regimes". *Agriculture, Ecosystems and Environment*, vol 93, no. 1, p. 147-164, 2002.

McClintock, J. T.; Schaffer, C. R.; Sjoblad, R. D. A comparative review of the mammalian toxicity of *Bacillus thuringiensis*-based pesticides, *Pesticide Science*, vol. 45, no. 2, pp. 95–105, 1995.

Meadows, M. P. D.; Ellis, J.; Butt, J.; Jarrett, P.; Burges, H. D. Distribution, frequency and diversity of *Bacillus thuringiensis* in an animal feed mill. *Applied Environmental Microbiology*, vol. 58, p. 1344-1350, 1992.

Monnerat, R. G.; Silva, S. F.; Silva-Wenerck, J. O. Catalogo do Banco de germoplasma de bactérias entommopatogênicas do gênero *Bacillus*. *Embrapa Recursos Genéticos e Biotecnologia,* 65 pp, 2001.

Monnerat, R. G.; Dias, D.; Silva, S.; Martins, E.; berry, C.; Falcão, R.; Gomes, A. M. M.; Praça, L., Soares, C. M. Screening of *Bacillus thuringiensis* strains effective against mosquitoes. *Pesquisa em Agropecuária Brasileira*, vol. 40, p. 103-106, 2005.

Monnerat, R. G.; Batista, A. C.; Mederios, P.; Martins, E.; Melatti, V. M.; Praça, L. Screening of Brazilian *Bacillus thuringiensis* isolates active against *Spodoptera frugiperda, Plutella xylostella* and *Anticarsia gemmantalis*. *Biological Control*, vol. 41, p. 291-295, 2007.

MONSANTO. Evaluation of the new active Bacillus thuringiensis var. kurstaki delta-endotoxins as produced by the 1Ac and 2 Ab genes and their controlling sequences in the new product Bollgard® II cotton event 15985. Available at http://www.monsanto com.au/content/ cotton/bollgard _ii_cotton/publicrelease.pdf. 2003.

Morse, R. J.; Yamamoto, T.; Stroud, R. M. Structure of 2Aa suggests an unexpected receptor binding epitope. *Structure,* vol. 9, no. 5, p. 409–417, 2001.

Nakasu, E. Y. T.; Firmino, A. A. P.; Dias, S. C.; Rocha, T. L.; Ramos, H. B.; Oliveira, G. R.; Lucena, W. A.; Carlini, C. R.; Grossi de As, M. F. Analysis of 8Ka5-binding proteins from *Anthonomus grandis* (Coleoptera: Curculionidae) midgut. *Journal of Invertebrate Pathology,* vo. 104, p. 227-230, 2010.

Oliveira-Filho, E. C.; Oliveira, R. S.; Lopes, M. C.; Ramos, F. R.; Grisolia, C. K.; Alves, R. T.; Monnerat, R. G. Toxicity assessment and clearance of brazilian microbial pest control agentes in Mice. *Bull Environmental Contaminant and Toxicology,* vol. 83, p. 570-574, 2009.

Oliveira, G. R.; Silva, M. C. M.; Lucena, W. A.; Nakasu, E. Y. T.; Firmino, A. A. P.; Beneventi, M. A.; Souza, D. S. L.; Gomes Jr., J. E.; Souza Jr.; J. D. A.; Rigden, D. J.; Ramos, H. B.; Soccol, S. R.; Grossi-de-Sa, M. F. Improving 8Ka toxin activity towards the cotton boll weevil (*Anthonomus grandis*). *BMC Biotechnology*, p. 11-85, 2011.

Oppert, B.; Kramer, K. J.; Johnson, D. E.; MacIntosh, S. C.; McGaughey, W. H. Altered protoxin activation by midgut enzymes from a *Bacillus thuringiensis* resistant strain of Plodia interpunctella. *Biochemistry and Biophysics Research Commun.*, vol. 198, p. 940–947, 1994.

Pacheco, S.; Gómez, I.; Arenas, I.; Saab-Rincon, G.; Rodríguez-Almazán, C.; Gill, S. S.; Bravo, A.; Soberón, M. Domain II loop 3 of *Bacillus thuringiensis* 1Ab tpxin is involved in a "pingo pong" binding mechanism with *Manduca sexta* aminopeptidase-N and cadherin receptors. *Journal of Biological Chemistry*, vol. 284, p. 32750-32757, 2009.

Park, Y.; Abdullah, M. A. F.; Taylor, M. D., Rahman, K.; Adang, M. J. Enhancement of *Bacillus thuringiensis* 3Aa and 3Bb toxicities to coleopteran larvae by a toxin-binding fragment of an insect cadherin. *Applied Environmental Microbiology*, vol. 75, p. 3086–3092, 2009.

Piggot, C. R.; Ellar, D. J. Role of receptors in *Bacillus thuringiensis* stal toxin activity. *Microbiology and Molecular Biology Reviews,* vol. 71, p. 255-281, 2007.

Poncet, S.; Anello, G.; Delècluse, A.; Klier, A.; Rapoport, G. Role of the IVD polypeptide in the overall toxicity of *Bacillus thuringiensis* subsp. *israelensis. Applied Environment Microbiology*, vol. 59, p. 3928-3930, 1993.

Poncet, S.; Delécluse, A.; Klier, A.; Rapoport, G. Evaluation of synergistic interactions among the IVA, IVB, and IVD toxic components of B.

thuringiensis subsp. Israelensis stals. *Journal of Invertebrate Pathology,* vol. 66, p. 131–135, 1995.

Pretty, J. N.; Brett, C.; Gee, D.; Hine, R. E.; Mason, C. F.; Moison, J. I. L.; Raven, H.; raiment, R. D.; Bijl, G. An assessment of the total external costs of UK agriculture. *Agricultural Systems,* vol. 65, no. 2, p. 113-136, 2000.

Puntheeranurak, T.; Uawithya, P.; Potvin, L.; Angsuthanasombat, C.; Schwartz, J. L. Ion channels formed in planar lipid bilayers by the dipteran-specific 4B Bacillus thuringiensis toxin and its a1-a5 fragment. *Molecular Membrane Biology,* vol. 21,p. 67–74, 2004.

Qaim, M.; Zilberman, D. Yield effects of genetically modified crops in developing countries. *Science,* vol. 299, no. 5608, p. 900–902, 2003.

van Rie, J.; Jansens, S.; Hofte, H.; Degheele, D.; van Mellaert, H. Receptors on the brush border membrane of the insect midgut as determinants of the specificity of *Bacillus thuringiensis* delta-endotoxins. *Applied Environmental Microbiology,* vol. 56, p. 1378–1385, 1990.

Rochester, I. J. Arthropod Management, Effect of genotype, edaphic, environmental conditions, and agronomic practices on 1Ac protein expression in transgenic cotton. *The Journal of Cotton Science,* vol. 10, p. 252–262, 2006.

Schrøder, M.; Poulsen, M.; Wilcks, A. A 90-day safety study of genetically modified rice expressing 1Ab protein (*Bacillus thuringiensis* toxin) in Wistar rats, *Food and Chemical Toxicology,* vol. 45, no. 3, p. 339–349, 2007.

Schuler, T. H.; Poppy, G. M.; Kerry, B. R.; Denholm, I. Insect-resistant transgenic plants. *Trends in Biotechnology,* vol. 16, p. 169-175, 1998.

Sims, S. R.; Ream, J. E. Soil inactivation of the *Bacillus thuringiensis* subsp. *kurstaki* IIA insecticidal protein within transgenic cotton tissue: laboratory and field studies. *Journal of Agriculture and Food Chemistry,* vol. 45, p. 1502-1505, 1997.

Smith, R.; Couche, G. The phylloplane as a source of *Bacillus thuringiensis* variants. *Applied Environmental Microbiology,* vol. 57, p. 311-315, 1991.

Soberón, M.; Pardo, L.; Muñoz-Garay, C.; Sánchez, J.; Gómez, I.; Porta, H.; Bravo, A. Pore formation by toxins. In: Anderluh, G.; Lakey, J. H. (Eds.), *Proteins: membrane binding and pore formation.* Springer, New York, p. 127-138, 2010.

Tailor, R.; Tippett, J.; Gibb, G.; Pells, S.; Pike, D.; Jordan, L.; Ely, S. Identification and characterization of a novel *Bacillus thuringiensis* delta-endotoxin entomocidal to Coleopteran and Lepidopteran larvae. *Molecular Microbiology,* vol. 6, p. 1211-1217, 1992.

Trisyono, A.; Whalon, M. E. Fitness costs of resistance to Bacillus thuringiensis in Colorado potato beetle (Coleoptera: Chrysomelidae). *Journal of Economic Entomology,* vol. 90, p. 267–271, 1997.

U. S. Environmental Protection Agency. Biopesticides Registration Action Documents *Bacillus thuringiensis* Plant-Incorporated Protectants. *Office of Pesticide Programs,* 2001. Retrived from http://www.epa.gov/ pesticides/biopesticides/pips/bt_brad2/3-ecologi-cal.pdf.

Vachon, V.; Laprade, R.; Schwartz, J-L. Current models of the mode of action of *Bacillus thuringiensis* insecticidal stal proteins: A critical review. *Journal of Invertebrate Pathology,* vol. 111, p. 1-12, 2012.

Vidal-Quist, J. C.; Castañera, P.; González-Cabrera, J. Diversity of *Bacillus thuringiensis* strains isolated from *Citrus orchads* in Spain and evaluation of their insecticidal activity against *Ceratitis capitata. Journal of Microbiological Biotechnology,* vol. 19, no. 8, p. 749-759, 2009.

Xiao, L.; Liu, C.; Xie, C. C.; Cai, J.; Chen, Y. H. The direct repeat sequence upstream of *Bacillus thuringiensis,* genes is cis-acting elements that negatively regulate heterologous expression in *E. coli. Enzyme Microbiology and Technology,* vol. 50, no. 6-7, p. 280-286, 2012.

Whalon, M. E.; Wingerd, B. A. Bt: mode of action and use. *Archives of Insect Biochemistry and Physiology,* vol. 54, p. 200-211, 2003.

Weinzierl, R.; Henn, T.; Koehler, P. G.; Tucker, C. L. Microbial Insecticides. *Office of Agricultural Entomology, University of Illinois at Urbana-Champaign.* Entomology and Nematology Department, Florida Cooperative Extension Service, Institute of Food and Agricultural Sciences, University of Florida. First published: June 1995. Revised: June 2005.

Wright, D. J.; Iqbal, M.; Granero, F.; Ferre, J. A change in a single midgut receptor in the diamondback moth (*Plutella xylostella*) is only in part responsible for the field resistance to *Bacillus thuringiensis* subsp. *kurstaki* and *B. thuringiensis* subsp. *aizawai. Applied Environmental Microbiology,* vol. 63, p. 1814–1819, 1997.

Zhang, J.; Fitz-James, P. C.; Aronson, A. I. Cloning and characterization of a cluster of genes encoding polypeptides present in the insoluble fraction of the spore coat of *Bacillus subtilis. Journal of Bacteriology,* vol. 175, no. 12, p. 3757-3766, 1993.

Zhang, X.; Candas, M.; Griko, N. B.; Rose-Young, L.; Bulla, L. A. Cytotoxicity of Bacillus thuringiensis 1Ab toxin depends on specific binding of the toxin to the cadherin receptor BT-R1 expressed in insect cells. *Cell Death Differentiation*, vol. 12, p. 1407–1416, 2005.

Zhang, X.; Candas, M.; Griko, N. B.; Taussing, R.; Bulla, L. A. A mechanism of cell death involving an adenylyl cyclase/PKA signaling pathway is induced by the 1Ab toxin of *Bacillus thuringiensis. Procedures of the National Academy of Sciences USA,* vol. 103, p. 9897-9902, 2006.

In: Macromolecular Chemistry: New Research   ISBN: 978-1-62417-854-2
Editor: Valentin Gartner   © 2013 Nova Science Publishers, Inc.

Chapter 2

STRUCTURING OF THE POLYMER COMPOSITES WITH CHEMICALLY ACTIVE FILLERS

Iu. Bolbukh[*]

Chuiko Institute of Surface Chemistry, National Academy of Science of
Ukraine, Kyiv, Ukraine

ABSTRACT

At the present the main tasks in nanotechnology development are to combine the desired properties of the filler and polymer; to identify the factors which are determine the structure-properties relationship, as well as reinforcement effects of the filler embedded into the polymeric matrix. Nevertheless, a unique interaction between components within the composite structure complicates systematization. Despite of the considerable number of researches, the factors which determine strengthening of the final composite are unclear. Comprehensive study of the interface and dynamic effects during filler-filler and filler-polymer interaction will intensify understanding of the strengthening mechanism of the polymer based composites.

The modern theories of the polymer materials strengthening and factors which determine the structural effect of the silica fillers are discussed in this work. Influence of the filler surface modification on the polymer based composite structure is presented. Polymer matrices, filled

---

[*] yu_bolbukh@yahoo.com.

with nanoparticles with mono and bifunctional grafted layer is in focus of this work.

It is show that the properties of the composite are conditioned by characteristics of the polymer layer near and on the surface of filler particles. A determinative at forming of composite structure is ability of the filler surface directionally to orient the growing in the process of polymerization macromolecules, and in case of co-polymers - to change also the composition of polymeric blocks due to selective adsorption of components of the monomers mixture. This work demonstrates that the final properties of the composite are conditioned by the structure of the polymer layer near and on the surface of filler particles. Ability of the filler to arrange monomers during polymerization is a key factor at the forming of composite structure; such arrangement is possible due to the selective adsorption of the initial component mixture.

INTRODUCTION

Rapid development of the nanotechnology leads to diversifying of the material types with new functional properties. Significant fractions of this material are filled polymers and nanosized silica's are commonly used fillers [1].

Adjusting in the dispersity of silica fillers, as well as particles shape and the chemistry of their surface leads to control their properties.

At the present stage of the nanotechnology development main tasks in the synthesis of composites are to establish principles of creating materials with a combination of desired properties both the filler and the polymer, as well as to identify the factors that determine the structure-properties relationship and the reinforcing effect of the filler. Interaction of the fillers and the polymeric matrix resulting in the desirable structure-properties of the final composite is the primary goal of the nanocomposite science. However, the components interaction is unique and complicates systematisation. Despite the numerous research [2, 3], the question about determining factors of the composites reinforcement is still controversial.

Comprehensive study of the filler-filler and filler-polymer interface during polymer-filler interaction should lead to deep understanding in the reinforcement mechanism of the final composite.

This work analyze and summaries the current state of the science as well as review the recent literature in order to revise the different theories of composite strengthening in applying to the effect of the filler surface functionality on the polymer structure.

1. THE THEORIES OF COMPOSITE STRENGTHENING

Experimental data shows that the reinforcing effects of the filler depends on the content of filler in the system, dispersity of the filler [2], size and shape of the particles [4, 5], the filler specific surface area, and on the nature of the polymer-filler interaction [6, 7]. The main theories of the composite strengthening are summarized in Table 1. In spite of the differences in key factors all theories are converge that the composite strengthening is achieved due to an interaction at the polymer-filler interface and it is caused by change in the properties of the polymer at the interface.

Indisputably, the filler dispersity and mutual arrangement of the macromolecules and the inorganic nanoparticles of the filler within the composite structure is an additional factor in structure reinforcement [8, 9, 10]. At the same time, the mobility and surface activity of the filler particles are important. Under these conditions, a composite reinforcement is strongly dependent on individual characteristics of both polymer and filler [1].

Most popular theory for reinforcement of the polymers with inorganic fibers was the theory of chemical bound. With an transition to nano dimensions of filler this theory is still popular, but more relevant theories are consider composites as the three-phases system (there and further the phases in three-phases systems are polymer, filler and intermediate) whereas reinforcement is defined by polymer properties on the filler surface. Determining factors are associated with the volume of filler fractions, particle size distribution and the thickness of polymer layer on interface. These parameters are interlinked. Overall, with decreasing in size of filler particle the effect of third phase (intermediate) increases, but an increase in thickness of the polymer layer on the filler surface leads to improved mechanical properties of the material only when the size of the filler particles less than 50 nm [15]; polymer in the composites pass into the plastic state with decreasing in the particle size of filler. Distribution and concentration of the local tensions are depends on structure and volume of the third phase; the reinforcement in material, whereas properties of the filler and polymer are contrary, is achieved through the polymer layer on the filler surface.

The «third phase» characteristic can be changed through modification of the filler surface, which leads to the certain structurization in the phase. Combination of the entropy and enthalpy effects improves the control in morphology of composite [19]. Moreover, an interaction between polymer and the functionalized filler is ensuring the specific localization of the particle in matrix. Under this condition, the architecture of the modifying layer, nature of interaction and activity of the attached functional groups are play a significant role. It should be noted that the composite reinforcement is depends not only on the strength of the interaction in interphase, but also on the flexibility of polymer chains [20, 21].

Table 1. The theories of composite strengthening

Theory	Main statements	Ref.
Surface wettability theory	Polymer phase that forms a solvation shell around the filler particles has higher mechanical properties and ensures the continuity of the material. The main determining factor of composite reinforcing is filler's wettability by the polymer (or monomer).	[11]
Theory of deformable (appreted additive) layer	A layer of appreted additive relieves the local stresses on the filler surface during polymer curing that provides an increase in the strength of material.	[12]
Theory of restrained layer	Appreted additive forms the compact (restrained) polymer structure in the polymer-filler surface layer.	[13]
Chemical bonding theory	Composite strengthening is provided by the interaction between functional groups on filler particles and polymer chains with formation of chemical bonds. The determining factor is a filler-polymer crosslink density. Formation of the chemical bounds between polymer chains and the functional groups on filler surface results in the composite strengthening; key factor is a filler-polymer crosslink density.	[14]
Theory of the third phase.	Polymer layer around the filler particles is considered as the "third phase" and mechanical characteristics of the material depends on the characteristics these third phase. Key factors are volume of the filler, filler's particle size and depth of the polymer at the interface.	[15].
Theory of the network geometry	Key factor in the material reinforcement is the volume fraction of the polymer on the filler surface and the geometry of network of the filler particles enveloped in polymer. Interaction in this case conducted through the polymer shells. In regards to this theory gradient of the polymer properties on the surface of filler and in the composite volume conditioning the composite properties.	[16]
The theory of the reinforcement compound	At the absence of the chemical and/or significant adsorption interaction between polymer and inorganic particles, the mechanism of polymer materials reinforcement is conducted through the formation of the reinforcement structure from the filler particles. Utilizing of the fillers which are able to form cross-linked network with significant coupling in the aggregates and to form a bridging bonds with macromolecules prevents degradation of the polymer. Structure and specific surface area of the filler play a significant role, because adsorbed macromolecules are more stable at the material destruction.	[17]
The theory of the interpenetrating network	The separated filler aggregates are had no significant influence on the general composite reinforcement, therefore the polymer-filler interpenetrating network improving the mechanical properties of the composite. Depending on the filling degree this structure is composed from spherical formations, external layer of which is composed from polymer, structured by filler; core is non-structured polymer or non-structured polymer locked in-between structured phase of polymer-filler interpenetrating network.	[18]

Depending on the functionality of the organic group, attached to silica, the filler becomes hydrophobic, or the organo-silica is able to form covalent bond with polymer macromolecules. Accepted that formation of the three-dimensional network via covalent bonding between phases is improves the mechanical properties and thermostability of the composite. Among such structures the effective properties can be specific through individual interaction polymer/filler [22, 23]. Whereas the presence of the organic groups in the surface layer of silica is prevents the phase immiscibility, it does not ensure the maximal reinforcement in composite.

It is coursed by competition of several processes:

- compaction of the surface layer with it thermostability increase;
- formation of cross-linked network of the polymer-filler bonds or silica particles framework;
- loosening the surface layer due to increased rigidity of macromolecules;
- formation overstress bonds with the forming of the regions with different cross linking density [24].

In this case, determining factors are the strength of the polymer / filler interaction (and thus the nature and activity of the filler surface), which in each composite have the optimum obtained only by experiment, and the influence of the filler surface on a polymerization mechanism.

2. EFFECT OF FILLER SURFACE ON POLYMERIZATION PROCESS

During forming of a composite structure an action of the filler surface depends on polymerisation type and method of curing. At radical polymerization of vinyl monomers in the presence of initiators alive polymeric chains are stabilised by the silica surface that reduces the chain termination rate [25]. An increase of the polymerisation rate under such conditions is explained by catalytic action of silica surface on a process of the initiator decomposition, as well as by an orientation and an ordering of monomer molecules on a solid surface [26]. The orientation and ordering of monomer molecules, even in the absence on the surface centres of the polymerization, facilitates the process [27]. At a radiation curing of the compositions based vinyl monomers (methyl methacrylate, styrene, etc.) the active polymerization centres are $\equiv$Si-OH group of the filler surface. A radiation-chemical decomposition of these groups is accompanied by initiation of the polymer radicals' growth. However, the surface hydroxyls could to cut down the polymer chains, as observed in radiation-initiated radical polymerization of polymethylmethacrylate [28]. In the presence of grafted chemically active groups a filler surface action is determined by surface group functionality and by features of filler-polymer interaction including mutual compatibility, wettability, dispersibility. So, grafted silicon hydride groups on the silica surface can promote both the acceleration and slowing the polymerizations of 2-hydroxyethylmethacrylate (HEMA) depending on the surface modification degree (content of the silicon hydride groups) as well on the type of polymerization reaction [29]. At the same time, the vinyl groups on the silica surface accelerate the radical polymerization of acrylates, regardless of the degree of filler surface modification. Therefore, the strategy in the synthesis of functionalized silica-filled polymer materials is accurate selection of functional groups as for surface hydrophobisation as well as modification with chemically active sites.

3. EFFECT OF FILLER SURFACE ON COMPOSITE STRUCTURE

As already noted, the properties polymers (thermal and mechanical) can be improved by the introduction of inert or chemically active fillers, namely, fillers with functional surface groups that form covalent bonds with the polymer. In the first case, properties of the material cause by only physical factors, such as the mobility of macromolecules in the surface layer or directing influence of the filler surface. For chemically active fillers a contribution of physical and chemical factors depends on reactivity of the grafted surface groups

Nanosized silica particles as known can be localized on the interface and promote the stabilization of polymer domains, their morphology and size [23, 30]. This improves the overall mechanical continuity of the material [31]. Contrary to this ability nanosized particles prone to aggregation. A control filler particles dispergation can be achieved by selection of functional groups for surface modification [32, 22]. However, modification of the silica surface reduces the adsorption ability and prevents the particles to form a stable coagulation network. Surface functionalization allows to obtain desired a filler wettability by dispersion medium and to provide a polymer / filler interaction [3, 32, 33].

Consider an influence of some functional groups on the structure of polymer composites. Surface modification of the filler by the alkenes is providing an affinity of components in the mixture; straightening in material is ensured by ability of the filler to hold an adsorbed polymer (monomer) on interface and pre-form gel state in disperse media. Ability to structurization in alkyl-silicas depends on the degree of the surface modification and length of the grafted radical.

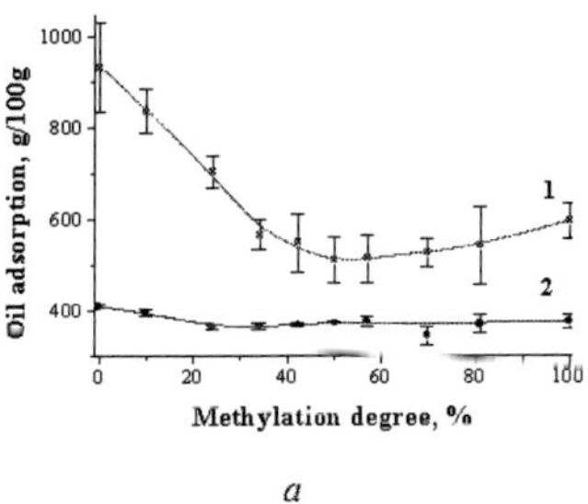

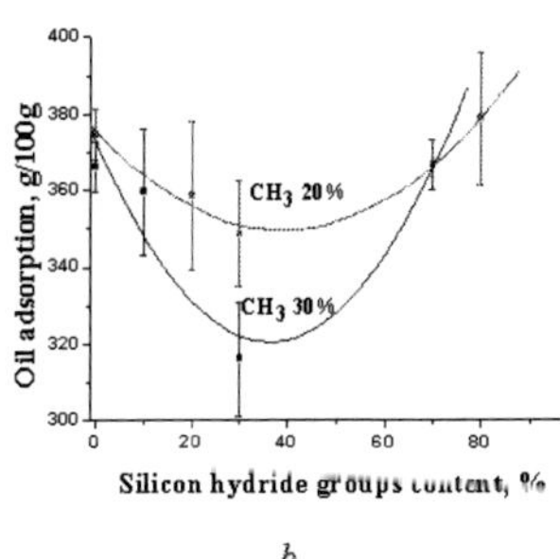

Figure 1. Oil adsorption of silica fillers: (a) - in various modification degree with $\equiv$Si(CH$_3$)$_3$ groups without (1) and after treatment with ethanol (2), (b) silica with the degree of methylation of 20 and 30% and different content of grafted silicon hydride groups [35].

For example, the gel state in polyethylene glycol is achieved by adding of the silica with 50% octyl modification; ether 50% dimethyl modification results in sol microstructure [34]. Hydrogen bonds and dispersion interaction in filler are leads to formation of the filler particles coagulates which results in media structurization. With an increase in degree of the silica modification up to hydrophilic-hydrophobic transition an ability of the filler to structurize the disperse media decrease (Figure 1, a, curve 1), such decrease was evaluated from values of oil-adsorption [35]. This can prove a significant role of the hydroxyl groups in structurization. Modification of the surface for 50% and above enlarge the contribution of the disperse interaction between attached organic groups. Composite structure is defined by contribution of the different type of interaction. During the introducing of the silica in monomer (polymer) the shape and size of silica agglomerates is altering; this change an ability of silica to structurize the liquid media. Thus, after treatment of the modified silica's with ethanol the values of oil-adsorption are depends negligible on the modification degree (Figure *1*, *a*, curve 2).

According to [36] the complete substitution of the surface hydroxyls by the organic groups leads to significant decline in the composite mechanical properties. It is attributed to distortion of the filler coagulation network. However, the filler coagulation network existence is not ensuring the composite reinforcement Thermostability of the composites of divinylbenzene (DVB) with di(methacrylo-methoxy)naphthalene (DMN) is decreasing equally with filling by 20% methylated silica as well as fully methylated silica [33]. An introducing of the alkyl-modified filler leads to the impairment in composite thermostability for majority of polymers. The surface activity of methylsilica's is conditioned by the fact that methylated particles are more mobile in organic media, comparing to the particles of hydrophilic silica. This activity appears in change in the polymer microstructure. An addition of the 30% methyl-modified silica to the DVB-DMN co-polymer is leads to the high-crystalline composite (Figure 2.).

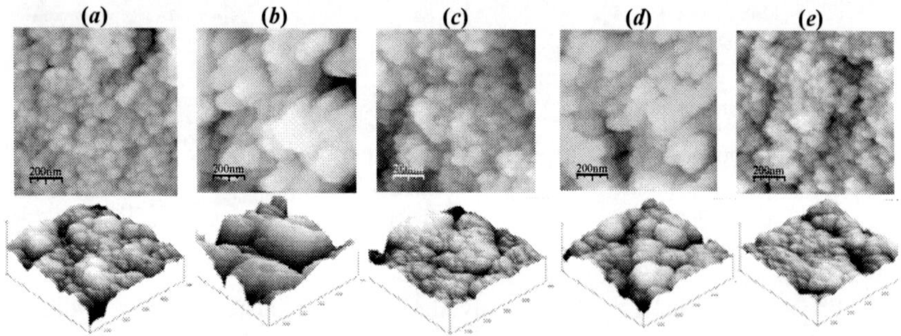

Figure 2. AFM micrograph of the surface of polymer spheres from DMN-DVB synthesized without (*a*) and in the presence of silica fillers with different surface functionality: *b* - 30%-Si(CH$_3$)$_3$ groups; *c* - the 20%-Si(CH$_3$)$_3$ and 30% ≡SiH groups; *d* - 30% C = C groups (Liquid-phase modification); *e* - 30% DVB grafted molecules.

The principal factor, in regards to current theories [20, 21] of the nanoparticle influence on the polymer morphology, is the energy of polymer/particle interaction. Dispersion interaction between matrix and methylated silica filler is ensure relative mobility in nanoparticles, those accumulates at the interface of weakly and strongly cross-linked areas and providing the direct growth of the macromolecules and densification in pre-formed domains [37].

Change in methylation degree is changing the sorption activity of the surface against the mixture compounds, and in turn the composition and orientation of macromolecules, as well as speed of relaxation in the overstrained bond. Thus, the dependence of the composite characteristics (thermostability, porosity, swelling, etc.) from the degree of surface methylation is complicated. Extremum in the analyzed characteristics is recorded at change in the surface lyophilic behavior [37]. And, the rate of polymerization process decreases with increasing degree of filler modification. Thus, a tension in structure caused by adsorption is weakened by slowing polymerization process. Compression of the macromolecules packing with an increase the filler modification degree is promoted by a relaxation of rigid (with a limited degree of mobility) units. Thus, the thesis about growing rigidity of the macromolecules with loosening in composite structure at filling is not always confirmed. In all probability that for methylated silica determine factor in the material strengthening is not only the filler ability to hold the polymer near the surface, but the ability of the surface to orient a growth of macromolecules in a certain way and to change theirs composition by selective adsorption of organic components. [33, 39]

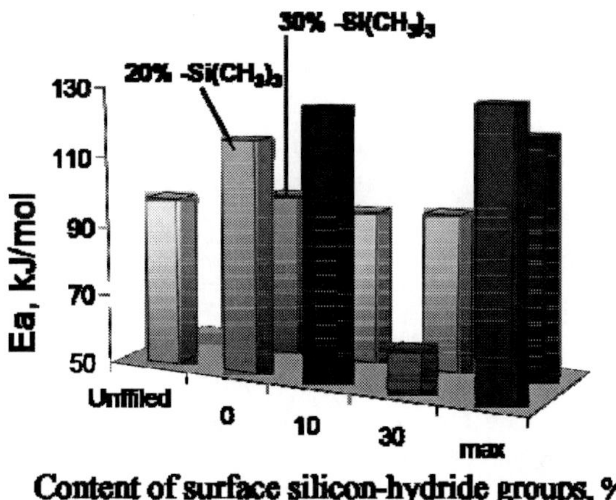

Figure 3. The activation energy of the thermal-oxidation of the composites based on DVB-DMN and silicas with varying modification degree by methyl-and silicon-hydride groups.

The materials based on modified silicas and copolymer DVB-DMNT were investigated in order to establish an influence of a functionality of the filler surface layer (namely, the concentration and ratio of chemically active and oleophilic groups, type and size of chemically active groups) on the structure and properties of polymer composites [38]. The silicas with bifunctional surface layer (methyl and silicon-hydride groups in different ratio), with grafted vinyl-groups and with chemosorbed divinylbenzene were used as fillers. The structure of the materials obtained was studied using a thermal analysis and by nitrogen ads/desorption method [37, 39].

According to results obtained, the presence on the filler surface both the methylsilil and the silicon-hydride groups ensures physical and chemical components of the polymer-filler interaction. The contribution of each component can be changed by varying the concentration of these groups. In the case of bifunctional surface layer the silicon hydride groups provide chemical bonding of macromolecules to the filler surface, and methyl groups - a relaxation of polymer chains at the polymer–filler interface, as evidenced by the results of thermogravimetric analysis (TGA) and differential scanning calorimetry (DSC) [33]. In Figure 3 shows the values of an activation energy of termal-oxidative degradation for composites filled with silicas containing on surface methylsilil and silicon hydride groups in different ratio. In general, the thermal stability of the material increases with increasing content of reactive groups. The highest thermal stability has the composite filled with silica with fully modified surface. Consequently, with increasing concentration of filler-polymer covalent bonds the material strain, that should occur under these conditions, relaxes due to the presence of methyl groups.

In addition to organic molecules orientation the composite strengthening depends on direction of polymerization (from or to surface of the filler), that is reflected in porosity of materials. The structure features of composites filled with methylated on 20%silica, containing 10-30% silicon hydride groups, testify an incipiency of copolymer chains on the surface with propagation in mixture volume (polymerization 'from the surface"). In these conditions, a significant loosening of composite structures is not observed (Figure 4). Despite of an increase of strained bonds content caused by increase a concentration of chemically active silicon hydride groups the system relaxation is occur. The polymerisation deceleration, in this case, assists forming a porous structure with relatively narrow pore size distribution (mostly d=25 nm). Given the small changes in the distribution of pores with increasing content of the silicon hydride groups, we can assume that methyl groups of the surface have a decisive role in pores formation.

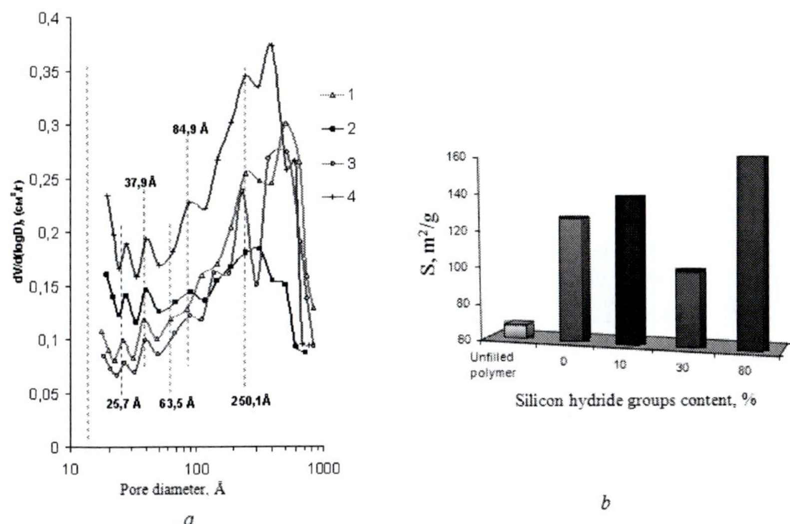

Figure 4. The pore size distribution (a) and specific surface area (b) for composites based on copolymer DVB-DMN and silica with varying degrees of surface modification by silicon hydride and methylsilyl groups *1* - 20% -Si(CH$_3$)$_3$ groups; *2* - 20% -Si(CH$_3$)$_3$ and 10% ≡SiH groups; *3*- 20% -Si(CH$_3$)$_3$ and 30% ≡SiH groups; *4* - 20% -Si(CH$_3$)$_3$ and 80% ≡SiH groups.

For composites filled with silica containing methyl and silicon hydride groups as 20 and 80%, respectively, an incipiency of copolymer chains occurs in the material bulk and terminates on silica surface (polymerization "to the surface"). At polymerization "to the surface" the pores volume increases but the pore size distribution does not change significantly. A high specific surface area caused by formation a loosening structure with multi-level pore network due to internal skeleton composed by polymer chains which is chemically cross-linked through an interaction of filler silicon hydride groups and vinyl groups of monomers. In general, the surface silicon hydride groups ensures the formation internal skeleton from polymer-filler bonds which changes the content of pores with diameter near and above 250 Å , while the concentration of methylsilil groups determines overall pore size distribution.

An increase concentration of silicon hydride groups, with increasing porosity and cross-linking degree of the composite, promotes the formation more flexible, but stable bonds. However, it is confirmed only for composites with hydride containing silicas methylated on 20%. A small change in the ratio of methyl and silicon hydride groups significantly changes the properties of the composite (Figure 3, *5*) [37].

In contrast to composites with 20% methylated hydride-containing silicas which ensure a growth flexibility of macromolecules with packaging compressing and reducing the specific surface area of the materials, filling with silicas methylated on 30% results in high specific surface area and pore volume of materials obtained (Figure 5). At using silicas with 30% degree of surface methylation an increase the silicon hydride groups concentration don't lead to significant change of the material porosity. As for the 20% methylated hydride-containing silicas, for silicas with 30% methylation degree the direction of polymerization depends on the chemically active (silicon hydride) groups content. But in this case, when concentration of $\equiv$ SiH groups in the range 10-30% the polymerisation passes "to the surface", and at 70% - "from the surface." These changes are caused by formation on the silica surface with 30% methylation degree the silicon-containing layer with another architecture, in compare to 20% methylated silica. Perhaps, during modification with triethoxysilane the dimers or oligomers of alkoxy-silanes have been immobilized on silica surface. Obviously, in bifunctional surface layer the chemically active groups ensure an orientation of the macromolecules. In this case, the macrostructure of copolymer is maintained (Figure 2). In case of bifunctional layer the effect of the filler surface on polymer chain growing, propagation and termination become more significant, but the determining factor is precisely the orientation of macromolecules in the filler surface layer [39-41].

Vinyl groups grafted to the silica surface, in contrast to silicon hydride, are capable significantly improve the surface hydrophobicity without the presence of methyl groups. As the filler with silicon hydride groups, vinyl modified filler provides composite strengthening. Moreover, in the presence of vinyl groups on the surface the fillers are more active in process of macromolecules forming [33]. It should be note that activity of surface groups depends on the method of silica modification (gas or liquid-phase modification), consequently, on the environment of active groups and on the size of filler particles aggregates: the fillers obtained by gas-phase modification are more active in polymerization process. Also, at 30% gas-phase modification with vinylalkoxysilane the polymer-filler interaction are realized by the mechanism "from the surface," and at 60% - "to the surface." In the case of liquid-phase surface modification, the polymerization of monomers carried out from the surface, regardless of the content of grafted groups.

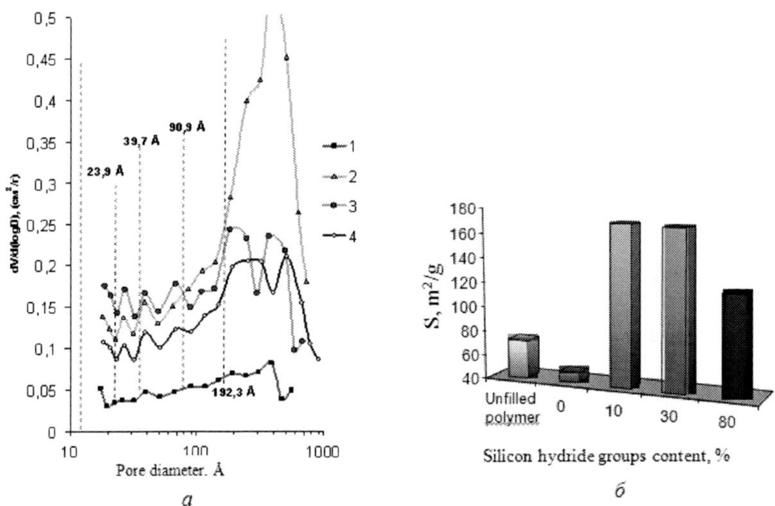

Figure 5. The pore size distribution (a) and specific surface area (b) for composites based on copolymer DVB-DMN and silica with varying degrees of surface modification by silicon hydride and methylsilyl groups: *1*- 30% -Si(CH$_3$)$_3$ groups; *2* - 30% -Si(CH$_3$)$_3$ and 10% ≡SiH groups ; *3*- 30% -Si(CH$_3$)$_3$ and 30% ≡SiH groups ; *4* - 30% -Si(CH$_3$)$_3$ and 80% ≡SiH groups.

High activity in polymerization reaction of the vinyl-containing fille at 100% surface modification leads to a decrease in activation energy of termooxidative degradation (Figure 6), and hence - to loosening of the surface layer structure. However, in contrast to metyl-hydride modified silicas, vinyl-containing filler ensures the formation of porous structure with a relatively narrow pore size distribution (Figure 7) [40]. An increase a concentration of grafted vinyl groups promotes improving of the parameters of material porosity with signs of multi-level pore network (Figure 7, *a*). Thus, with the full covering of native silica surface with hydrophobic groups and at the presence in the surface layer residual ethoxy-group the structure of the polymer surface layer is loosened. It should be noted that the orientation of the macromolecules, which causes the pore size distribution, depends on the filler agglomerates size [41] and type of a modifier.

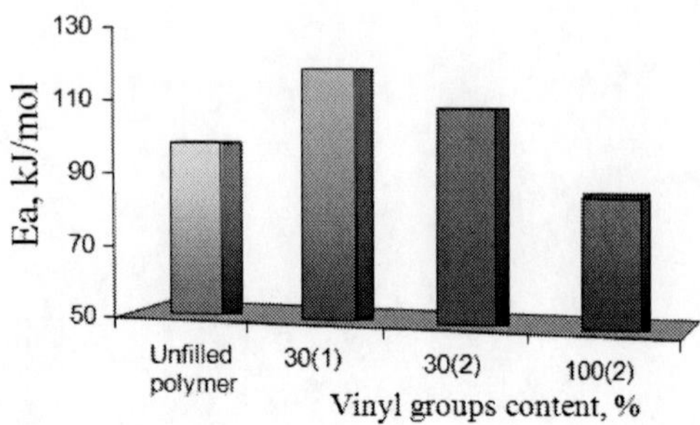

Figure 6. The activation energy of the thermooxidative degradation for composites based on copolymer DVB-DMN and silicas with different degree of surface functionalisation with vinyl groups grafted via gas-phase (1) or liquid-phase modification (2).

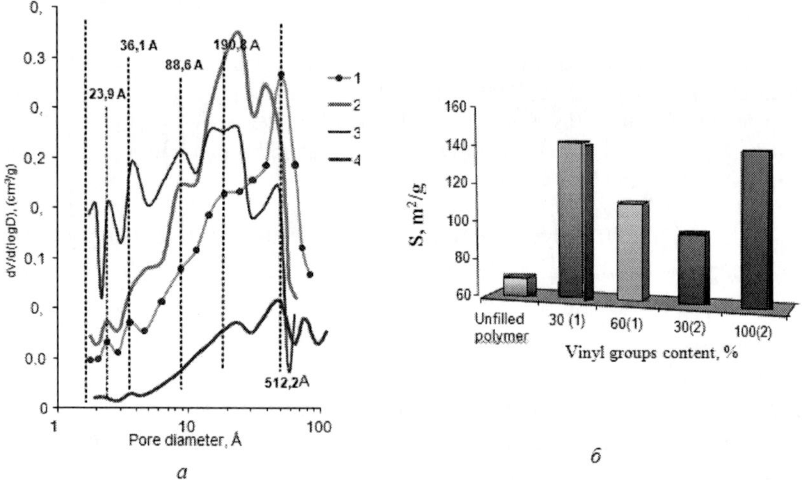

Figure 7. The pore size distribution (*a*) and specific surface area for composites (*b*) based on copolymer DVB-DMN and silicas with varying degree of surface modification with vinyl-groups: *1* - unfilled DVB-DMN; *2* - 30% -Si(CH=CH$_2$) groups (gas-phase modification); *3* - 60% -Si(CH=CH$_2$) groups (gas-phase modification); *4*- 30% -Si(CH=CH$_2$) groups (liquid-phase modification); *5*- 100% -Si(CH=CH$_2$) groups (liquid-phase modification).

Orientation of macromolecules also depends on the size of grafted functional groups and their location in the surface layer. The filled polymers porosity can be controlled by elongation of a distance between functional groups and surface that prevents a compression of the surface layer. Analysis the obtained results for copolymer DVB-DMN with the silica modified by divinylbenzene shows that 30% of filler surface coating provides a significant increase in material porosity. However, there is an optimum of modification degree and the complete coverage of the surface with grafted molecules leads to disturbance of structure (Figure 8). At total surface modification a material discontinuity is marked: a macromolecules packing in the surface layer is compacted with structure loosening in bulk [37, 40, 41]. At the same time, a compressing of packing in surface layer stabilizes the drops of monomer mixture in the dispersion media, contributing to the structure with small domens of macromolecules, similar to the structure of unfilled copolymer, but with greater dispersion (Figure 2). The interaction of the copolymer and filler modified with divinylbenzene is limited by steric factor that increases with modification degree increasing. In the case of 30% degree of modification the steric factor is practically absent.

The composite filled with silica totally modified (100% surface covering with divinylbenzene) showed unexpectedly low temperature of thermooxidative degradation (Figure 8, *b*), however, for this material there are two degradation stages with a significant difference between the activation energies of the processes.

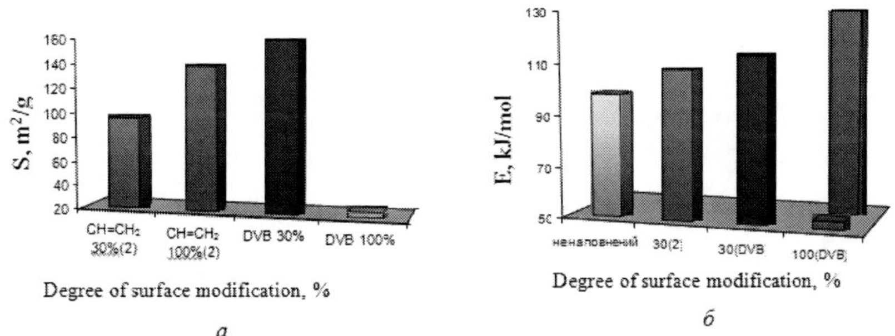

Figure 8. Specific surface area (a) and activation energy of thermooxidation (b) for composites based on DVB-DMN and silicas with different degree of surface modification with vinylalkoxysilane or divinylbenzene.

In general, vinyl groups of the modifier grafted promote chain transfer during polymerization. An increase a concentration of grafted groups leads to increase heterogeneity of the material structure. Enlarging the size of the functional group and distance from the filler surface leads to increase rigidity of copolymer chains and to reducing the size of polymer domains. At the same activity attached group a significant role plays the size of filler particles aggregates. The groups with smaller size are more suitable for a homogeneous structure forming.

CONCLUSION

In order to establish structure-property relationships within existing theories it is necessary to study the influence of the functional groups of the filler on the structure of the polymer in surface layer of inorganic particles. Controlling of this structure is achieved by forming on silica filler surface the layer with certain architecture and activity. The determining factor in system under study is the surface ability to direct the growing macromolecules in certain way, and, in the case of copolymers, change the composition of the copolymer blocks due to selective adsorption. Therefore, to obtain the desired properties of the material an investigation should be directed to the study of a surface adsorption activity towards system components and the degree of layer ordering and orientation of the monomer (polymer) molecules during formation of the composite. A direction of the polymerization (from or to filler surface) and surface activity of the filler, which provides formation of the certain size of polymer domains, are important for synthesis of composite with desired structure.

REFERENCES

[1] Zou H., Wu S., and Shen J. / Polymer/silica nanocomposites: preparation, characterization, properties, and applications// *Chem. Rev.* – 2008. –V. 108. – P.3893-3957.

[2] Mahrholz T., Stangle J., Sinapius M. / Quantitation of the reinforcement effect of silica nanoparticles in epoxy resins used in liquid composite molding processes // *Composites, A.* – 2009. - V. 40. – P. 235–243.

[3] Smith J.S., Bedrov D., Smith G.D. / A molecular dynamics simulation study of nanoparticle interactions in a model polymer–nanoparticle composite // *Compos. Sci. Tech.* - 2003. – V. 63. – P. 1599–1605.

[4]   Zhou Q., Xanthos M. / Nanosize and microsize clay effects on the kinetics of the thermal degradation of polylactides // *Polym. Degrad. Stab.* – 2009. – V. 94. – P. 327–338.

[5]   Siengchin S., Karger-Kocsis J. / Structure and creep response of toughened and nanoreinforced polyamides produced via the latex route: Effect of nanofiller type // *Compos. Sci. Tech.* – 2009. – V. 69. – P. 677–683.

[6]   Vansant E.F., Van Der Voort P., and Vrancken K.C. *Characterization and Chemical Modification of the Silica Surface.*- Amsterdam: Elsevier, 1995. - 560 p.

[7]   Lipatov Y.S. / Polymer blend and interpenetration polymer networks at the interface with solids // *Prog. Polym. Sci.* – 2002. – V. 27. – N 9. – P. 1721-1801

[8]   Plueddemann E.P. / *Composite materials.* /L.J. Broutman and R.H. Krock (Eds.)/ - Academic Press: New York and London. – 1974. - V. 6. - P. 181-227.

[9]   Synergistic effect of carbon black and nanoclay fillers in styrene butadiene rubber matrix: Development of dual structure / S. Praveen, P.K. Chattopadhyay, P. Albert, V.G. Dalvi, B.C. Chakraborty, S. Chattopadhyay // *Composites, A.* – 2009. – V. 40. – P. 309–316.

[10] Mechanical properties of experimental dental composites containing a combination of mesoporous and nonporous spherical silica as fillers / S. P. Samuel, S. Li, I. Mukherjee, Y. Guo, A. C. Patel, G. Baran, Y. Wei // *Dent. Mater.* – 2009. – V. 25. – P. 296–301.

[11] Zisman W.A. / Influence of constitution on adhesion // *Ind. Eng. Chem.* – 1963. – V. 55. – N 10. – P. 18–38.

[12] Hooper R.C., *11th Ann. Techn. Conf. SPI, Reinforced Plastics. Div., Sect.* 8-B, 1956.

[13] Kumins C.A., Roteman J. / Effect of solid polymer interaction on transition temperature and diffusion coefficients // *J. Polym. Sci., A.*– 1963. – V. 1. – P. 527-540.

[14] Bjorksten J, Yaeger L.L. / Vinyl silane size for glass fabric // *Modern Plastics.* - 1952. – V. 29. – P. 124-188.

[15] Micromechanics-based modelling of stiffness and yield stress for silica/polymer nanocomposites / S. Boutaleb, F. Zairi, A. Mesbah, M. Nait-Abdelaziz, J.M. Gloaguen, T. Boukharouba, J.M. Lefebvre // *Int. J. Solids Struct.* – 2009. – V. 46. – P. 1716–1726.

[16] Qiao R., Brinson L. C. / Simulation of interphase percolation and gradients in polymer nanocomposites // *Compos. Sci. Tech.* – 2009. – V. 69. – P. 491–499.

[17] Kluppel M. / The role of disorder in filler reinforcement of elastomers on various length scales // *Adv. Polym. Sci.* – 2003. – V. 164. – P. 1–86.

[18] Reinforcement effects in nanocomposites enhanced by aggregation and percolation phenomena / P. Mele, S.Marseau, D.Brown, N.D.Alberola // *EuroFillers03, Alicante* (Spain), September 8-11, 2003.—P. 225-228.

[19] Balazs A.C., Emrick T., Russell T.P. / Nanoparticle polymer composites: Where two small worlds meet // *Science.*- 2006. – V. 314. – P. 1107-1110.

[20] Hooper J.B. and Schweizer K.S. / Theory of phase separation in polymer nanocomposites // *Macromolecules.* – 2006. – V. 39. – P. 5133-5142.

[21] Effect of nanoscopic particles on the mesophase structure of diblock copolymers / J.Y. Lee, R. B. Thompson, D. Jasnow, A.C. Balazs, // *Macromolecules.* - 2002. – V. 35. - P. 4855-4858.

[22] Novel precipitated silicas: an active filler of synthetic rubber / T. Jesionowski, A. Krysztafkiewicz, J. Zurawska, K. Bula // *J. Mater. Sci.* – 2009. – V. 44. – P. 759–769.

[23] Govsky D.Y. / Devise application of polymer-nanocomposites // *Adv. Polym. Sci.*—2000.—V. 153. — P. 163-205.

[24] Lipatov Yu.S. Polymer Reinforcement. - *ChemTech Publishing:* 1995. - 406 p.

[25] Reiko S., Koji K., Tadakuni T. / Synthesis of poly(methylmethacrylate)-silica nanocomposite // *J. Macromol. Sci.—Pure Appl. Chem.*— 2002. —V. 39A. - N 3.— P. 171-182.

[26] Tertykh V., BolbukhYu, Yanishpolskii V. / Effect of unmodified and surface treated fumed silica on the polymerization of HEMA // *Macromol. Symp.* – 2005. – V. 221. – N 1. - P 145 - 152.

[27] Kontou E., Anthoulis G. / The effect of silica nanoparticles on the thermomechanical properties of polystyrene // *J. Appl. Polym. Sci.* - 2007. – V. 105. – P. 1723–1731.

[28] Bryk M., *Polymerisation on solid surface.*— Kyiv: Naukova dumka, 1981.—288 p.

[29] Bolbukh Yu, Mamunya Ye.P., Tertykh V.A. / Thermomechanical investigation of composites based on 2-hydroxyethylmethacrylate and higt disperse initial and modified silica // *J. Therm. Anal. Cal.* - 2005.- V. 81. – P. 15-21.

[30] Vaia R.A. and Maguire J. F. / Polymer nanocomposites with prescribed morphology: going beyond nanoparticle-filled polymers// *Chem. Mater.* – 2007. – V. 19. – P. 2736-2751.

[31] Schmidt G., Malwitz M.M. / Properties of polymer–nanoparticle composites // *Curr. Opinion Colloid Interface Sci.* – 2003. – V. 8. – P. 103–108.

[32] Properties/structure relationships in innovative PCL–SiO$_2$ nanocomposites. / M. Avella, F. Bondioli, V. Cannillo, S. Cosco, M.E. Errico, A.M. Ferrari, B. Focher, M. Malinconico // *Macromol. Symp.* – 2004. – V. 218. – P. 201–210.

[33] Effect of surface hydride, vinyl, and methyl groups on thermal stability of modified silica-divinylbenzene-di(methacryloyloxymethyl) naphthalene composites / Bolbukh Y.M., Kuzema P.O., Tertykh V.A., Gawdzik, B. // *Int. J. Polym. Mater.* – 2007. - V. 56. - N 8. – P. 803 – 823.

[34] Composite polymer electrolytes based on poly(ethyleneglycol) and hydrophobic fumed silica: dynamic rheology and microstructure / S.R. Raghavan, M.W. Riley, P.S. Fedkiw, and S.A. Khan // *Chem. Mater.* – 1998. V. 10. - P. 244-251.

[35] Kuzema P., Bolbukh Yu. / High dispersed silicas with silicon hydride groups as fillers for polymers // Book of Abstracts. *"Tekhnology of XXI senchery".* – Alushta, 2006. – P. 98-102.

[36] Filler-filler interaction and filler-polymer interaction in carbon black and silica filled ExxproTM polymer / W.K. Wong., G. Ourieva, M.F. Tse, H.C. Wang // *Macromol. Symp.* – 2003. – V. 194. - P. 175-184.

[37] Bolbukh Iu.M., Tertykh V.A., Gawdzik B. / Preparation and characterization of porous polymer composites filled with chemically modified silica // *Polish J. Chem.* – 2008. – V.82. - P.17–24.

[38] Bolbukh Y., Tertykh V., Gawdzik B. / *Porous polymeric nanocomposites filled with chemically modified fumed silicas* / In: Surface Chemistry in Biomedical and Environmental Science, J.P. Blitz and V. Gun'ko (Eds.), NATO Science Series: II. Mathematics, Physics and Chemistry, Springer: Dordrecht. 2006. - V. 228. - P.103-111.

[39] Bolbukh Yu., Tertykh V., Gawdzik B. / DTA AND DSC studies of filled porous copolymers // *J. Therm. Anal. Cal.* - 2006.- V. 86. – N 1. - P. 125-132.

[40] Preparation and structure of porous polymer nanocomposites with chemically active silica fillers /Yu.M. Bolbukh, V.A. Tertykh, P.O. Kuzema, B. Gawdzik / *Proc. X Ukrainian- Polish Symp. on Theoretical*

and Experimental Studies of Interfacial Phenomena and their Technological Applications, Part 1, Lviv-Uzlissia, Ukraine, 26-30 September – 2006.- P.32.

[41] Tertykh V., Bolbukh Yu., Gawdzik B. / Effect of silica filler functionalization on structure of porous polymer spheres, Book of abst. *Europ. Polymer Congress*, 2007, July 1-6, 2007, Portoroz, Slovenia, P.57.

In: Macromolecular Chemistry: New Research   ISBN: 978-1-62417-854-2
Editor: Valentin Gartner   © 2013 Nova Science Publishers, Inc.

Chapter 3

ENRICHMENT OF SAUVIGNON BLANC WINE BY HEAT STABLE WINE MACROMOLECULES

Johannes de Bruijn[1], José Martínez-Oyanedel[2], Francisco Lobos[2], Constanza Anziani[1] and Cristina Loyola[1]*

[1]Department of Agroindustry, University of Concepción, Chillán, Chile
[2]Department of Biochemistry and Molecular Biology, University of Concepción, Concepción, Chile

ABSTRACT

Proteins and glycoproteins have shown to play a major role in heat induced haze formation in white wines despite of their low concentrations. The aim of this study was to improve thermal stability of white wine by enrichment with heat stable wine macromolecules. First, macromolecules from Sauvignon Blanc wine were separated and concentrated by sequential membrane fractionation, followed by the precipitation with acetone and then further purified by two-dimensional polyacrylamide gel electrophoresis. The main heat stable macromolecular fraction consisted of a glycosylated protein with molecular weight (MW) between 69 and 72 kDa and an isoelectric point of 3.25, which was identified by MALDI-TOF/TOF mass spectrometry as vacuolar grape

---

* E-mail: jdebruij@udec.cl.

invertase 1, GIN1. Using Concanavalin A affinity chromatography, high molecular weight compounds of 74 kDa and >250 kDa were retained, increasing the heat instability of wine from 48 NTU to 292 NTU. After desorption of these retained compounds followed by the enrichment of wine with glycoproteins having molecular weights of 70-100 kDa and >250 kDa, heat instability decreased by maximum 56%. However, this effect was mainly due to the desorption solution. Addition of grape invertase to Sauvignon Blanc would not improve wine stability.

Keywords: Wine proteins, wine glycoproteins, thermal stability, grape invertase, haze, purification

INTRODUCTION

One of the major problems for winemakers is haze formation in bottled white wines due to protein instability. The appearance of haziness in a bottle of wine is unattractive for the consumer, who will reject wines containing cloudy precipitates regardless of their gustatory characteristics (Ferreira *et al.*, 2002; Waters *et al.*, 2005). Slow denaturation of wine proteins, resulting from unfavorable storage conditions such as exposure to high temperatures, is thought to originate protein aggregation and flocculation into a hazy suspension and finally, the formation of precipitates (Chi *et al.*, 2003). Usually wine industry tries to stabilize white wines to prevent protein precipitation. Currently the best way to reduce the risk of protein precipitation is by bentonite fining (Lambri *et al.*, 2012). However, this treatment is not very selective and may affect adversely wine quality by inducing significant aroma losses and being detrimental to foaming properties (Lubbers *et al.*, 1996; Dambrouck *et al.*, 2005; De Bruijn *et al.*, 2009). Moreover, bentonite fining results in a significant wine loss and a negative environmental impact due to the use of diatomaceous earth as filter aid for bentonite removal (Ribéreau-Gayon *et al.*, 2006). Therefore, alternative methods of white wine stabilization have to be developed. Although the protein concentration in unfined white wines is relatively low, varying typically from 15 to 300 mg/L (Ferreira *et al.*, 2002; Waters *et al.*, 2005), these components influence the sensory properties of the final product by stabilizing foam in the case of sparkling wines (Dambrouck *et al.*, 2005) and by increasing aroma intensity of model white wines (Jones *et al.*, 2008). Wine proteins have molecular mass ranging from 9 to 66 kDa and isoelectric point ranging from 3 to 9 (Hsu and Heatherbell, 1987; Brissonnet and Maujean, 1993; Dawes *et al.*, 1994). Several authors have demonstrated the glycosylated nature of many wine proteins (Waters *et al.*, 1993; Yokotsuka *et al.*, 1994; Marchal *et al.*, 1996), where the glycosyl group enhanced their stability (Batista *et al.*, 2009). Most of the identified wine proteins are from grape origin and can be classified in two groups: i) defense-related proteins, e.g. chitinase and thaumatin-like proteins, induced in grape berries by pathogens or environmental stress; ii) proteins involved in sugar metabolism, i.e. vacuolar invertases (Cilindre *et al.*, 2008). Grape invertase represents at least 10-20% of the wine proteins (Dambrouck *et al.*, 2005). On the other hand, only a minor fraction of wine proteins has been derived from yeast and could be identified as glycoproteins particularly rich in mannose (Dupin *et al.*, 2000). These mannoproteins were located exclusively in the cell wall of *Saccharomyces cerevisiae* and released by yeast during autolysis (Dupin *et al.*, 2000; Feuillat, 2003). Mannoproteins, which are present in wines at 100 – 150 mg/L, are complex compounds, having a wide range of molecular weights, extending from 5-53 kDa (Gonçalves *et al.*, 2002; Doco *et al.*, 2003) to 252-560 kDa (Waters *et al.*, 1994; Gonçalves *et al.*, 2002). The low molecular weight fraction was a mannose heteropolymer with 2.5% of proteins and 97.5% of saccharides (87.5% mannose and 12.5% of rhamnose, arabinose and galactose) and the high molecular weight fraction was a mannose homopolymer with 10.3% of proteins and 89.7% of saccharides (Gonçalves *et al.*, 2002). Addition of 500 mg/L of macromolecules from the yeast cell wall was able to reduce heat induced protein haze formation in white wines (Waters *et al.*, 1993; Dupin *et al.*, 2000; Lomolino and Curioni, 2007). In particular an invertase fragment from *S. cerevisiae*, purified on a Concanavalin A Sepharose column and identified as N-glycosylated 31.8 kDa mannoprotein, had a heat – stabilizing effect on the proteins after its addition (250 mg/L) to white wine (Moine-Ledoux and Dubourdieu, 1999). Polysaccharides are the other main group of macromolecules in wines. They come from grape berries, yeasts, or other microbial flora. Grape polysaccharides originate from the cell wall of grape berries and include arabinans, arabinogalactans, arabinogalactan-proteins and rhamnogalacturonans (Doco *et al.*, 2003). Concentration of the main wine polysaccharides is in the range of 10 to 100 mg/L, where molecular weights vary between 10 and 240 kDa (Vernhet *et al.*, 1999). In particular arabinogalactan-proteins have interesting physicochemical properties because of their capacity to inhibit protein haze formation (Waters *et al.*, 1993).

The thermo-sensitivity differs between the proteins of white wines (Falconer *et al.*, 2010). Most studies on wine proteins have been focused on heat unstable proteins (Estcruelas *et al.*, 2009; Sauvage *et al.*, 2010; Marangon *et al.*, 2011). However, heat stable proteins are important contributors to wine quality and more research on them is needed. Therefore, the identification and purification of heat stable macromolecules, followed by the enrichment of white wines with these heat stable wine macromolecules may improve overall thermal stability and subsequently wine quality.

Recently, we were able to separate proteins from Sauvignon Blanc wine into heat stable and heat unstable fractions by using membrane fractionation (De Bruijn *et al.*, 2011a, b). Based on these results, we aimed to further identify and characterize these heat stable macromolecules and to improve thermal stability of white wine by its enrichment with these compounds.

MATERIALS AND METHODS

Materials

Unfined wines were elaborated by an industrial winery according to standard white wine technology, using Sauvignon Blanc grapes from two Chilean valleys (CBV, Casablanca and CV, Curicó). All reagents used in the present study were of analytical grade. Chemicals for the preparation of buffers and reagents were purchased from Sigma-Aldrich (St. Louis, MO) unless otherwise stated. Characterization and identification of heat stable macromolecules were done using wines from 2009, where the enrichment studies were performed using wines from 2010.

Sample Preparation

Suspended solids were removed from wines by settling at 4°C for 40 h, centrifugation at 1,440 × g, filtration (Whatman No. 42, Whatman, Maidstone, UK) and microfiltration using a 0.45 µm pore size membrane (HVLP04700, Millipore Co., Bedford, MA). Wines were bottled in glass jars and stored at 4°C in the dark prior to ultrafiltration and affinity chromatography. Part of the samples was concentrated by sequential fractionation using composite regenerated cellulose membranes with a molecular weight cut off of 300, 100, 30 and 10 kDa (De Bruijn *et al.*, 2011a). Retentate samples were lyophilized in the Speed Vac (Thermo Savant, Holbrook, NY) and stored in Eppendorf tubes at -20°C before further purification. Then each sample was resuspended in 2 ml of distilled water followed by the addition of 8 ml of cold acetone. After 60 min at -20°C, samples were centrifuged at 12,000 × g for 10 min at 4°C. After removal of supernatant, pellet was dried at room temperature and resuspended in 1 ml of distilled water prior to electrophoresis and mass spectrometry. The fractions of 10-30 kDa and 30-100 kDa of both CBV and CV wines were analyzed by sodium dodecyl sulfate – polyacrylamide gel electrophoresis (SDS-PAGE) under reducing and non reducing conditions using different staining procedures. Wine proteins of the 30-100 kDa fractions of both wines were purified by two-dimensional electrophoresis (2D-E); their identification was done by matrix-assisted laser desorption / ionization time-of-flight (MALDI-TOF/TOF) mass spectrometry.

Prior to wine fractionation by using affinity chromatography, the composition of the remaining wine samples was modified, increasing wine pH to 4 using NaOH and by adding $CaCl_2$ (2 mM) and $MnCl_2$ (2 mM). Glycoproteins and polysaccharides from CV wine were purified by Concanavalin-A affinity chromatography and reintroduced in the unmodified wine samples followed by the assessment of their heat induced haze performance and (glico)protein profiles.

SDS-Page

Sodium dodecyl sulfate – polyacrylamide gel electrophoresis was performed according to Laemmli (1970) in a Mini Protean-3 apparatus (Bio-Rad Laboratories Inc., Hercules, CA). Resuspended aliquots (10 µl) were mixed with 2.5 µl of buffer (139 mM Tris-HCl, 2.3% (w/v) sodium dodecyl sulfate, 2 M urea, 27.8% (v/v) glycerol and 0.006% (w/v) bromophenol blue). If electrophoresis was done under reducing conditions, 5% (v/v) β-mercaptoethanol had been added previously to the buffer. Samples were incubated at 37°C for 2 h and centrifuged at 10,000 × g for 5 min before loading on a stacking gel, containing 125 mM Tris-HCl, 3.75% (w/v) polyacrylamide, 0.125% (v/v) Temed and 0.08% (w/v) ammonium persulfate. Using a resolving gel (375 mM Tris-HCl, 12.5% (w/v) polyacrylamide, 0.06% (v/v) Temed and 0.05% (w/v) ammonium persulfate) and buffer solution (250 mM Tris-HCl and 1.92 M glycine), electrophoresis was run at 100 V for about 135 min, when the tracking dye bromophenol blue reached the bottom of the gel. Protein markers (Prestained SDS-PAGE Standards, Broad Range, ref. 161-0318, Bio-Rad, Hercules, CA) were loaded at each run and used as molecular weight (MW) standards. After migration, proteins were stained with 0.18% Coomassie Brilliant Blue R-250 in acetic acid / methanol / water (1:4:6) and de-stained in acetic acid / methanol / water (2:3:6). Glycoproteins were detected by periodic acid - Schiff (PAS) staining (Allen *et al.*, 1976). Gels were placed in the above mentioned de-staining solution for 10 min. After washing with distilled water, fixed gels reacted with a 1% periodic acid solution for 30 min, washed three times for 5 min each in distilled water and stained with 2% Schiff base for about 30 min. MWs of unknown molecules were calculated from the linear regression equation of log MW versus mobility of molecular weight standards.

2D Electrophoresis

First-dimension isoelectrofocusing (IEF) was performed using a gel containing 8 M urea, 4% (w/v) acrylamide, 2% (v/v) Triton X-100, 1.6% (v/v) Bio-Lyte 5/7 Ampholyte, 0.4% (v/v) Bio-Lyte 3/10 Ampholyte, 0.1% (v/v) Temed and 0.01% (w/v) ammonium persulfate according to manufacturer's instructions (Bio-Rad Laboratories Inc., Hercules, CA). Sample (25 µl) was loaded together with buffer (25 µl) containing 8 M urea, 2% (v/v) Triton X-100, 5% (v/v) β-mercaptoethanol, 1.6% (v/v) Bio-Lyte 5/7 Ampholyte and 0.4% (v/v) Bio-Lyte 3/10 Ampholyte in a Mini-Protean 2-D apparatus (Bio-Rad Laboratories Inc., Hercules, CA). The composition of the electrophoretic buffer was: 4 M urea, 0.8% (v/v) Bio-Lyte 5/7 Ampholyte, 0.2% (v/v) Bio-Lyte 3/10 Ampholyte and traces of bromophenol blue. A cathode buffer of 100 mM NaOH and an anode buffer of 10 mM H_3PO_4 were used. The IEF migration program was as follows: start of rehydration and equilibration at 200 V for 5 min, increasing until 300 V for 15 min followed by 400 V for 15 min; start of focusing at 500 V for 10 min and final running at 750 V for 3.5 h.

Second-dimension SDS-PAGE was done by transferring the IEF gel to an already prepared SDS-PAGE gel, where separation was achieved in a buffer containing 139 mM Tris-HCl, 2.3% (w/v) sodium dodecyl sulfate, 2 M urea, 0.006% (w/v) bromophenol blue and 27.8% (v/v) glycerol, at 100 V until bromophenol blue marker reached the bottom of the gel. The PageRuler[TM] Plus Prestained Protein Ladder (SM1811, Fermentas, Genesys Chile S.A., Santiago, Chile) was used as MW standard.

After completion of electrophoresis, the gel was stained for proteins using the colloidal Coomassie Brilliant Blue G-250 procedure. This involves fixation in 45% (v/v) methanol and 2% (v/v) acetic acid for 20 min, followed by staining in 34% (v/v) methanol, 17% (w/v) ammonium acetate and 0.5% (w/v) Coomassie Brilliant Blue G-250 for 18 h. Finally the gel was transferred to distilled water and its surface rinsed to remove any particulate stain prior to scanning.

Mass Spectrometry

Peptide mass fingerprinting of selected protein spots from the 2D-E gels was carried out by in-gel trypsin treatment (Sequencing Grade, Promega, Madison, WI) at 37°C overnight. Peptides were extracted from the gels by Varian's OMIX pipette tips using 60% acetonitrile in 0.2% trifluoroacetic acid (TFA), concentrated by vacuum drying and desalted using C18 reverse phase micro-columns. Peptide elution from the micro-column was performed directly into the mass spectrometer sample plate with 3 µl of matrix solution (saturated solution of α-cyano-4-hydroxycinnamic acid in 60% aqueous acetonitrile containing 0.2% TFA). Mass spectra of digestion mixtures were acquired in a 4800 MALDI-TOF/TOF Analyzer (Applied Biosystems Inc., Framingham, MA) using the reflectron mode with external calibration using a mixture of peptide standards (Applied Biosystems Inc., Framingham, MA). Collision-induced dissociation MS/MS experiments of selected peptides were performed. Proteins were identified using the National Centre for Biotechnology Information no redundant (NCBInr) database, searching with peptide m/z values by using the MASCOT software (Matrix Science, London, UK) with the following search parameters: (1) mono isotopic mass tolerance of 0.08 Da; (2) fragment mass tolerance of 0.25 Da; (3) methionine oxidation, as possible modification, and (4) one missed tryptic cleavage was allowed. Analyses were carried out at the Pasteur Institute (Montevideo, Uruguay).

Concanavalin-A (Con-A) Affinity Chromatography

Sample (500 ml) of modified CV wine was loaded at a flow rate of 0.7 ml/min onto a Concanavalin-A column (HiTrap Con-A, GE Healthcare Life Sciences, Amersham, UK), previously equilibrated with binding buffer (20 mM Tris-HCl, pH 7.4, containing 0.5 M NaCl, 1 mM $CaCl_2$ and 1 mM $MnCl_2$). For elution of the bound fractions, an elution buffer (20 mM Tris-HCl, pH 7.4, containing 0.5 M NaCl and 0.5 M methyl-α-D-mannopyranoside) was used at a flow rate of 0.7 ml/min. Eluted aliquots (25 ml) were collected and analyzed for total protein concentration by measuring absorbance at 280 nm on a Sunny UV-7804C spectrophotometer, the presence of specific proteins and glycoproteins by electrophoresis and heat induced haze.

Finally, unmodified wine (25 ml) was enriched with several eluted fractions (1.19 ml) coming from the Con-A affinity column after desorption stage and heat induced haze performance was assessed.

Heat Induced Haze

Haze stability of wine was determined by the heat test according to Moine-Ledoux and Dubourdieu (1999), with some modifications. Each wine sample (25 ml) was transferred into a test tube sealed with a screw cap and heated for 2 h at 80°C in a thermostat-controlled water bath, then held at 4°C for 2 h, and warmed at room temperature for 2 h. Haze formation was measured using a Hach 2100P turbidimeter (Hach Co., Loveland, CO), where turbidity was expressed in Nephelometric Turbidity Units (NTU). Haze stability was expressed as the difference of wine turbidity (ΔNTU) before and after heat treatment. Analyses were carried out in triplicate. Wines are considered stable if the difference of turbidity does not exceed 2 NTU.

RESULTS AND DISCUSSION

Identification of Heat Stable Wine Macromolecules

After several purification steps, macromolecules from the retentate fractions of 30-100 kDa were separated by 2D – electrophoresis in diverse protein spots according to their molecular weight and isoelectric point, where spots were mainly distributed on the acid side (pH 3 – 5,5) of the gels with molecular weights (MWs) ranging from 15 to 65 kDa (Figure 1). Similar results have been reported by Cilindre et al. (2007).

Gel bands 1 and 2 (Figure 1) with MW of 65 kDa, were cut out for mass spectrometry analysis. Previously, wine proteins with a MW of 69-72 kDa according to SDS-PAGE did not show any correlation with wine haziness; these compounds should be considered as heat stable (De Bruijn et al., 2011a). The protein of both gel bands was identified as vacuolar invertase 1, GIN1 from *Vitis vinifera* (Table 1), which corresponds to the preferential expressed form encoded by GIN1 in berries (Davies and Robinson, 1996).

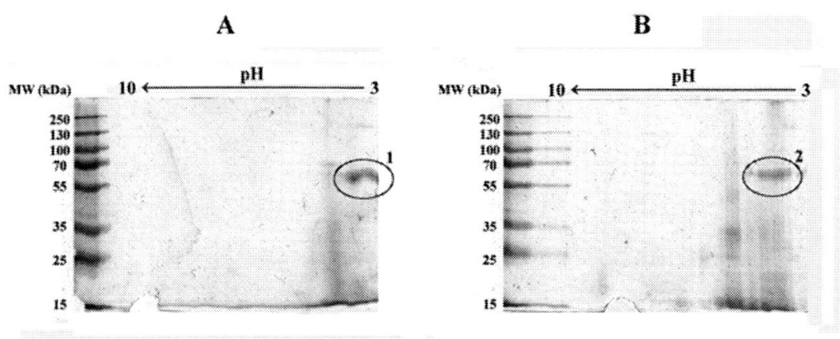

Figure 1. Two-dimensional electrophoresis of proteins from 30 – 100 kDa fractions of Casablanca (A) and Curicó (B) valley wines. Numbered spots were excised, digested by trypsin, and analyzed by MALDI – TOF / TOF mass spectrometry.

Sequence coverage of amino acids and peptide fingerprinting agree with literature data for vacuolar invertase 1, GIN1 depending upon previous deglycosylation and multiple protease digestions in the case of N-glycoproteins (Cilindre *et al.*, 2008; Jégou *et al.*, 2009).

However, no unique results about the thermo stability of vacuolar invertase 1, GIN1 from *Vitis vinifera* are reported in literature. Lambri *et al.* (2012) suggest that the denaturation of vacuolar invertase could be the main cause of haze formation observed in Sauvignon Blanc wine. Moreover, Esteruelas *et al.* (2009) detected an unidentified protein of 69 kDa in the natural haze protein of Sauvignon white wine. On the other hand, Marangon *et al.* (2011) confirmed the heat stability of a band at around 60 kDa that was assumed to be grape invertase. Vacuolar invertase 1 was heat stable until 60°C having a melting temperature of 81°C, where heating results into irreversible unfolding of invertase without evidence of protein aggregation (Falconer *et al.*, 2010; Sauvage *et al.*, 2010).

Characterization of Heat Stable Wine Macromolecules

In order to assess some properties of the heat stable vacuolar grape invertase 1, GIN1, some electrophoresis assays were performed. Because of the isoelectric point (pI) value of 3.25 (Figure 1), invertase is almost electrically neutral in Sauvignon Blanc with pH value of about 3.14, which is in contrast to most of the other wine proteins. Thus proteins with a more positive electrical charge seem to be more unstable.

Table 1. Identification of heat stable wine proteins by MALDI – TOF / TOF mass spectrometry purified by 2D electrophoresis

Band	Protein name (organism)	Accession number[a]	Number of peptides matched[b]	Sequence coverage[c]	Start - end	Peptide sequence
1	Vacuolar invertase 1, GIN1 (*Vitis vinifera*)	gi|1839578	7	9.5%	112-120 380-386 380-387 390-406 390-407 408-417 575-590	TAFHFQPEK TFYDQVK TFYDQVKK ILYGWISEGDIE SDDLK ILYGWISEGDIE SDDLKK GWASLQSIPR VLVDHSIVEGFS SQGGR
2	Vacuolar invertase 1, GIN1 (*Vitis vinifera*)	gi|1839578	8	7.3%	112-120 380-386 380-387 389-406 390-406 390-407 407-417 408-417	TAFHFQPEK TFYDQVK TFYDQVKK RILYGWISEGDI ESDDLK ILYGWISEGDIE SDDLK ILYGWISEGDIE SDDLKK KGWASLQSIPR KGWASLQSIPR

[a] Protein identification number provided by the NCBInr database.

[b] Number of peptides identified by MASCOT search engine that matched with the identified protein.

[c] Percentage ratio of all amino acids from valid peptides matched to the total number of amino acids in the reference proteins.

However, our experimental result is far from the theoretical pI value of 4.6 for this protein. Using two-dimensional polyacrylamide gel electrophoresis, Okuda *et al.* (2006) were able to detect three proteins with MWs of approximately 60 kDa having the same or very similar N-terminal amino acid sequences to that of invertase, showing pI values of 3.75 for the main spot and 4.23 and 5.16, respectively, for the minor spots.

Furthermore, Marchal *et al.* (1996) reported a pI of 3.9 for a 60/64 kDa protein, which should be the grape invertase. Variation between pI values could be due to micro heterogeneity of glycosylation, frequently observed among wine proteins.

Molecular weight is another characteristic of grape invertase that depends upon the method of biochemical analysis that has been used: MWs of 69-72 kDa for SDS-PAGE (Figure 2), 65 kDa using 2D-E (Figure 1) and 60 kDa for mass spectrometry were found in this study compared to the theoretical MW value of 71.5 kDa for vacuolar grape invertase 1, GIN1. This agrees with the difference of several thousands of Dalton found between MW of grape proteins in SDS-PAGE gels and MW determined by mass spectrometry (Pocock *et al.*, 2000), as well as the MW range reported before by Dambrouck *et al.* (2005).

SDS-PAGE analysis of soluble wine proteins in non-reducing conditions revealed ten bands with apparent MWs between 18 and 72 kDa (De Bruijn *et al.*, 2011b). However, different SDS-PAGE pattern was obtained under reducing conditions (Figure 2). In particular, the band at 21-23 kDa in the non-reducing samples was detected at about 27 kDa under reducing conditions (Figure 2).

This effect is probably due to the lower hydrodynamic volume of proteins when stabilized by intra-molecular disulfide bonds, where reducing conditions result into cleavage of disulfide bonds (Vincenzi *et al.*, 2011). On the other hand, grape invertase (69-72 kDa) maintained its place in both electrophoresis gels. These differences may be due to the number of intra-molecular disulfide bonds of the proteins.

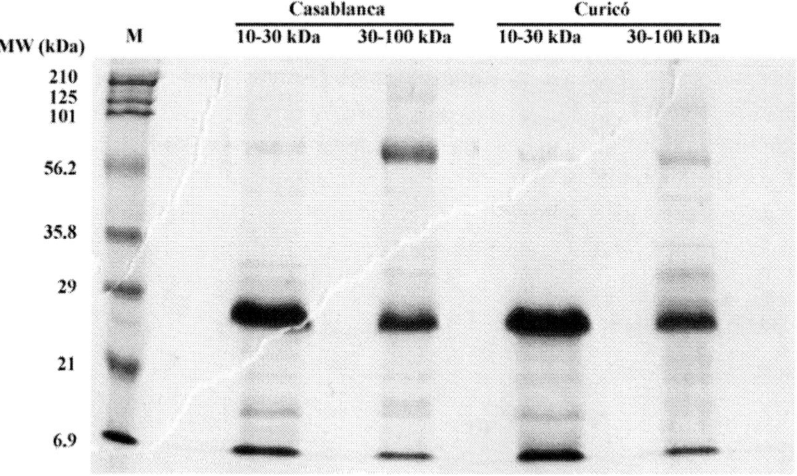

Figure 2. SDS-PAGE in reducing conditions after Coomassie Blue staining of proteins from 10 – 30 kDa and 30 – 100 kDa fractions of Casablanca and Curicó valley wines.

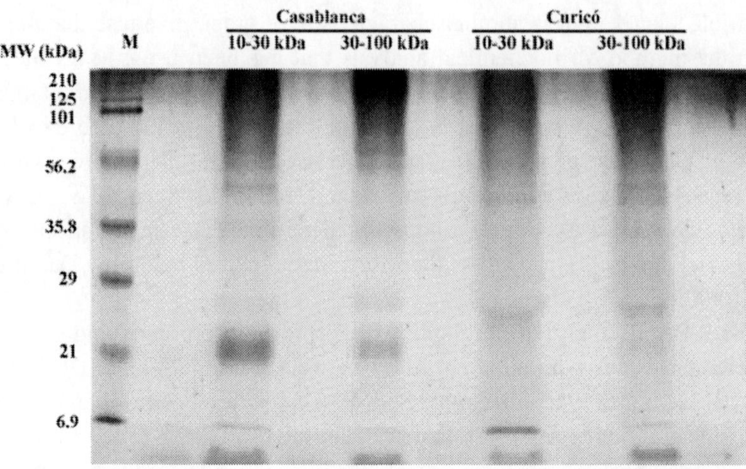

Figure 3. SDS-PAGE in reducing conditions after periodic acid - Schiff staining of glycoproteins from 10 – 30 kDa and 30 – 100 kDa fractions of Casablanca and Curicó valley wines.

The number of cysteine residues varies between sixteen for the thaumatin-like protein VVTL1 (MW of 24 kDa), fifteen for class IV endochitinase (MW of 28 kDa), fourteen for Grip22 precursor (MW of 23 kDa) and five for vacuolar invertase (*Vitis vinifera*) (MW of 72 kDa) (Lobos, 2011).

High molecular weight compounds (>60 kDa) including grape invertase, were stained by PAS after electrophoresis (Figure 3) and thus consist mainly of glycoproteins and polysaccharides. Indeed grape vacuolar invertase is an N-glycoprotein with twelve potential sites of N-glycosylation (Marchal *et al.*, 1996; Jégou *et al.*, 2009). Glycosylation of proteins increases their thermal and solution stability, and makes them less prone to proteolysis (Rowe *et al.*, 2010). The longevity of glycoproteins is related to folding constraints imposed by the glycosylation, where the carbohydrate moieties prevent unfolding or partially folded protein molecules from aggregation (Rowe *et al.*, 2010). Thus, the low degree of glycosylation of wine proteins having low molecular weight seems to be related to their low thermo stability.

Enrichment of White Wine

After about five bed volumes a sharp break – through of proteins and a strong increase of heat instability of eluted wine can be observed in Figure 4.

Specific retention of glycosylated proteins and polysaccharides from Sauvignon Blanc on the column of Con-A affinity chromatography seems to be responsible for the increased thermal instability of Sauvignon wine. Indeed after staining of the SDS-PAGE gel by periodic acid – Schiff reagent and by Coomassie Brilliant Blue R-250 the presence of only low MW proteins (19, 21 and 26 kDa) and glycoprotein (25 kDa) can be observed (Lane 1 of Figure 6 and Figure 7) in the highly unstable A1 fraction (Figure 4), where no high MW species neither proteins nor glycoproteins could be detected.

This indicates a preferential adsorption of higher MW carbohydrate species on the Con-A affinity column. After elution of about twelve bed volumes of wine the HiTrap Con-A system has been completely saturated, i.e. all active adsorption sites have been occupied by carbohydrate species. No further protein adsorption was found and turbidity difference of eluted aliquots agrees with the value of 47.6 ± 3.3 NTU of modified wine. Indeed the A2 fraction recollected at the end of the adsorption stage shows both glycoproteins and proteins in the range of 60 – 70 kDa (Lane 2 of Figure 6 and Figure 7), which agrees with the protein and glycoprotein profile of modified wine (Lane 5 of Figure 6 and Figure 7).

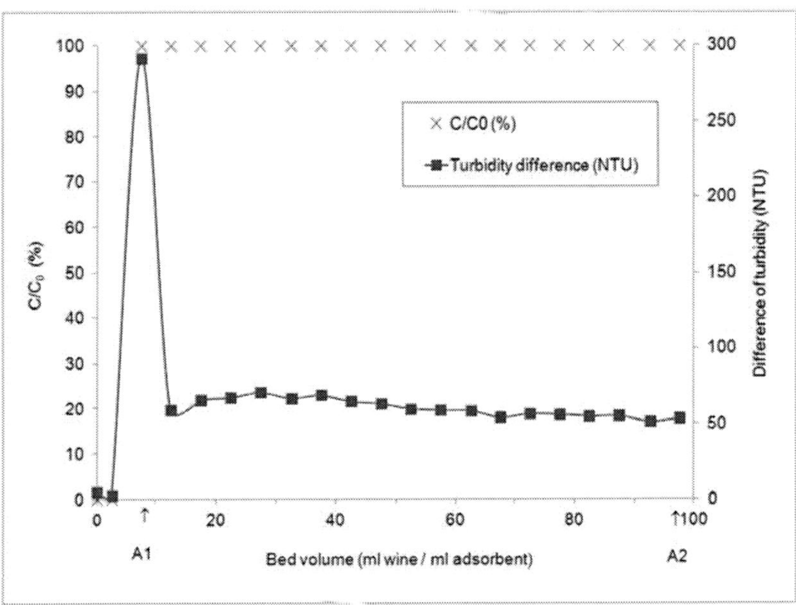

Figure 4. Break – through curves of total protein content and turbidity difference of Sauvignon Blanc wine using a HiTrap Concanavalin-A column.

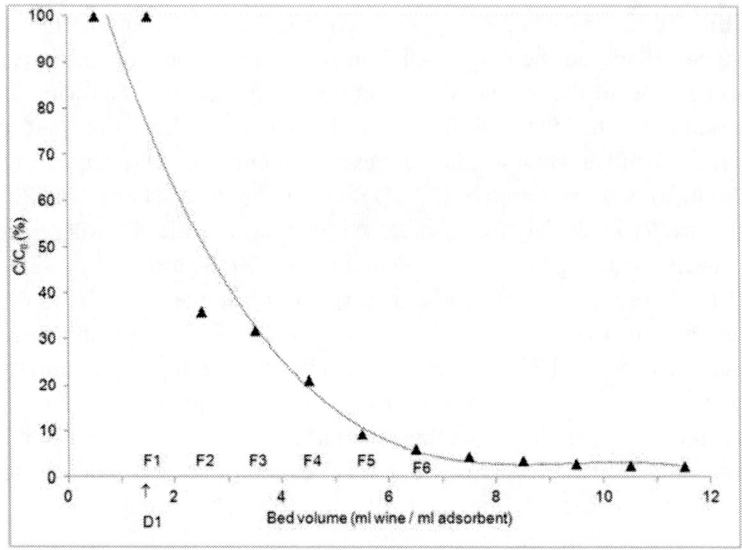

Figure 5. Regeneration of HiTrap Concanavalin-A column.

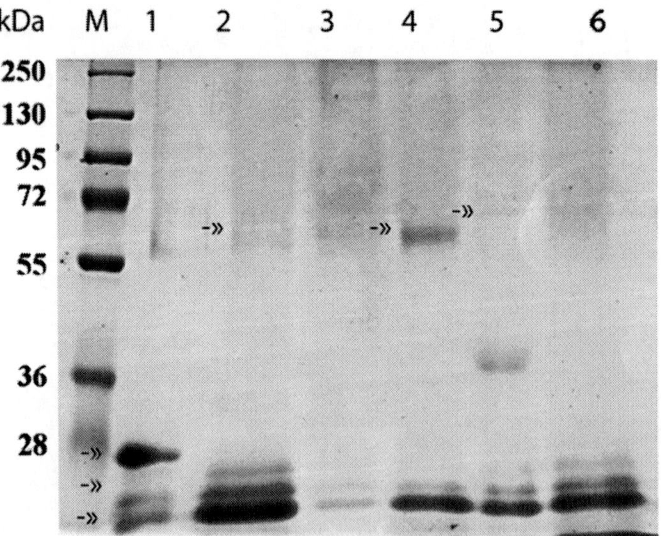

Figure 6. SDS-PAGE analyses of Curicó valley wine protein fractions after Coomassie Blue staining. Lane M, molecular weight standard proteins (kDa); lane 1, A1 fraction; lane 2, A2 fraction; lane 3, D1 fraction; lane 4, unmodified wine; lane 5, modified wine; lane 6, enriched wine with D1 fraction.

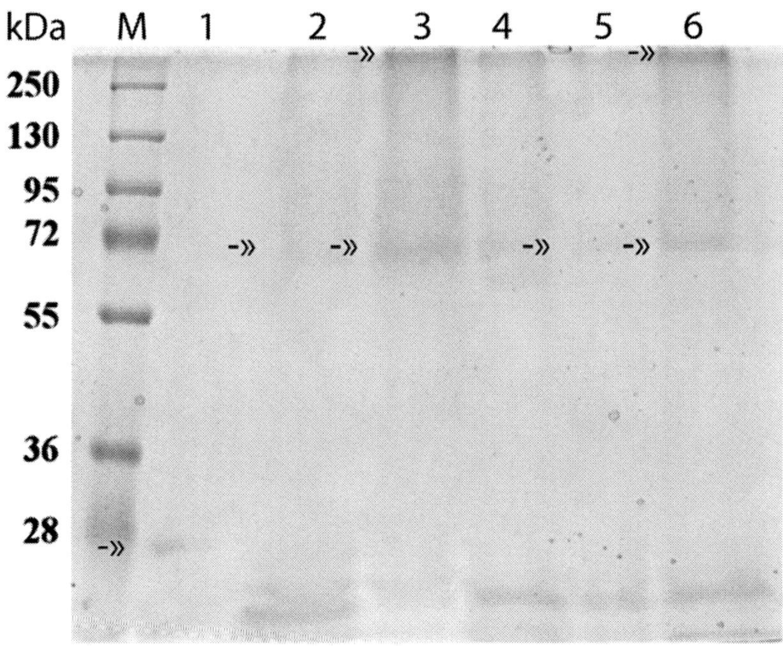

Figure 7. SDS-PAGE analyses of Curicó valley wine glycoprotein fractions after periodic acid - Schiff staining. Lane M, molecular weight standard proteins (kDa); lane 1, A1 fraction; lane 2, A2 fraction; lane 3, D1 fraction; lane 4, unmodified wine; lane 5, modified wine; lane 6, enriched wine with D1 fraction.

During the desorption stage retained species were removed by using about eight bed volumes of elution buffer with methyl-α-D-mannopyranoside (Figure 5). The first aliquot, the D1 fraction (Figure 5), was assessed for proteins and glycoproteins, showing main presence of high MW species (glycoproteins and/or polysaccharides) ranging between 70 kDa and >250 kDa, as well as an almost absence of low MW proteins (Lane 3 of Figure 6 and Figure 7). The glycoprotein of 70 kDa is assumed to be the heat stable grape invertase, as previously identified and characterized.

Several aliquots (F1 to F6) from desorption stage were added to unmodified wine and heat-induced haze of enriched wine samples was assessed. Although a considerable improvement of 29-56% of thermo stability of enriched Sauvignon Blanc wine was measured (Figure 8), this seems to be due to the composition of elution buffer, which results into an improvement of 55% of heat induced stability, than the effect of addition of glycoproteins, such as grape invertase amongst others, and polysaccharides.

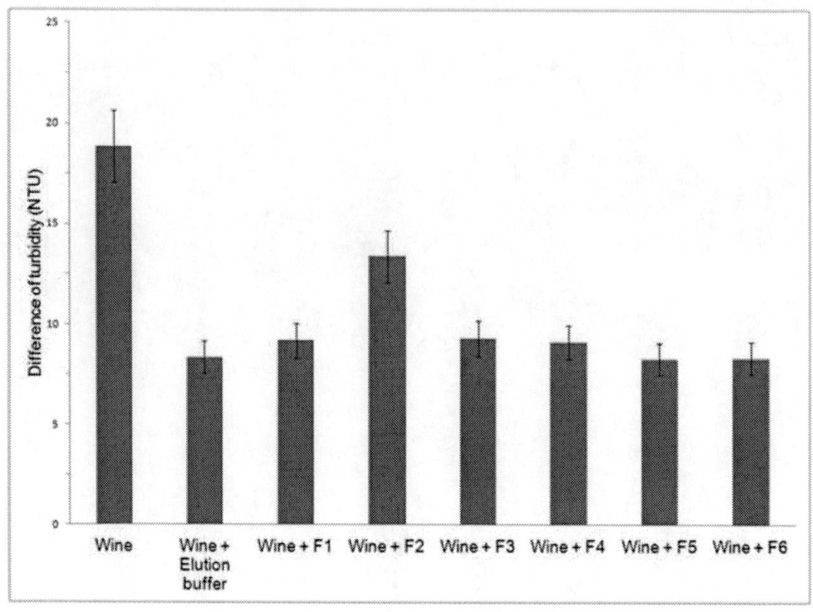

Figure 8. Thermo stability of wines. Unmodified wine; wine with elution buffer; wine enriched with F1, F2, F3, F4, F5 and F6 fractions from HiTrap Concanavalin-A column.

Moine-Ledoux and Dubourdieu (1999) reported that after purification on a column of Con-A Sepharose the addition (250 mg/l) of the retained fraction to wine was able to completely stabilize wine without haze formation after heating. They attribute the heat-stabilizing effect on the proteins in white wine to an N-glycosylated 31.8 kDa mannoprotein, which is an invertase fragment from *S. cerevisiae*. In our study, no significant amount of low MW glycoproteins in the aliquots after desorption stage could be observed.

CONCLUSION

The heat stable protein from Sauvignon Blanc wine is the vacuolar invertase 1, GIN1 from *Vitis vinifera* with molecular weight of 69-72 kDa and isoelectric point of 3.25, being glycosylated, which may contribute to the increased thermo stability of grape invertase compared to other wine proteins. Presence of glycoproteins and / or polysaccharides ranging between 70 kDa and >250 kDa, including heat stable grape invertase, in wine, is related to the improved heat induced stability of Sauvignon Blanc. The considerable improvement of thermo stability of wine that has been found after enrichment with heat stable, retained glycoproteins and polysaccharides from the HiTrap Concanavalin-A column was mainly due to the composition of the elution buffer and no positive effect of glycoproteins and polysaccharides could be detected.

ACKNOWLEDGMENT

We thank Mrs. Madelon Portela and Mrs. Rosario Duran for technical assistance. We are grateful to CONICYT (FONDECYT project No.11085054) for financial support.

REFERENCES

Allen, R. C. Spicer, S. S. and Zehr, D. (1976). Concanavalin A – horseradish peroxidase bridge staining of α-1 glycoproteins separated by isoelectric focusing on polyacrylamide gel. *Journal of Histochemistry and Cytochemistry, 24*, 908-914.

Batista, L. Monteiro, S. Loureiro, V. B. Teixeira, A. R. and Ferreira, R. B. (2009). The complexity of protein haze formation in wines. *Food Chemistry, 112*, 169-177.

Brissonnet, F. and Maujean, A. (1993). Characterization of foaming proteins in a champagne base wine. *American Journal of Enology and Viticulture, 44*, 297-301.

Chi, E. Y. Krishnan, S. Randolph, T. W. and Carpenter, J. F. (2003). Physical stability of proteins in aqueous solution: Mechanism and driving forces in nonnative protein aggregation. *Pharmaceutical Research, 20*, 1325-1336.

Cilindre, C. Castro, A. J. Clément, C. Jeandet, P. and Marchal, R. (2007). Influence of *Botrytis cinerea* infection on Champagne wine proteins (characterized by two-dimensional electrophoresis / immunodetection) and wine foaming properties. *Food Chemistry, 103*, 139-149.

Cilindre, C. Jégou, S. Hovasse, A. Schaeffer, C. Castro, A. J. Clément, C. Van Dorsselaer, A. Jeandet, P. and Marchal, R. (2008). Proteomic approach to identify champagne wine proteins as modified by *Botrytis cinerea* infection. *Journal of Proteomic Research, 7*, 1199-1208.

Dambrouck, T. Marchal, R. Cilindre, C. Parmentier, M. and Jeandet, P. (2005). Determination of the grape invertase content (using PTA-ELISA) following various fining treatments versus changes in the total protein content of wine. Relationships with wine foamability. *Journal of Agricultural and Food Chemistry*, 53, 8782-8789.

Davies, C. and S P Robinson, S. P. (1996). Sugar accumulation in grape berries. Cloning of two putative vacuolar invertase cDNAs and their expression in grapevine tissues. *Plant Physiology*, 111, 275–283.

Dawes, H. Boyes, S. Keene, J. and Heatherbell, D. (1994). Protein instability of wines: Influence of protein isoelectric point. *American Journal of Enology and Viticulture*, 45, 319-326.

De Bruijn, J. Loyola, C. Flores, A. Hevia, F. Melín, P. and Serra, I. (2009). Protein stabilisation of Chardonnay wine using trisacryl and bentonite: A comparative study. *International Journal of Food Science and Technology*, 44, 330-336.

De Bruijn, J. Martínez-Oyanedel, J. Loyola, C. Seiter, J. Lobos, F. and Pérez-Arias, R. (2011a). Fractionation of Sauvignon wine macromolecules by ultrafiltration and diafiltration: Impact of protein composition on white wine haze stability. *International Journal of Food Science and Technology*, 46, 1691-1698.

De Bruijn, J. Martínez-Oyanedel, J. Loyola, C. Lobos, F. Seiter, J. and Pérez, R. (2011b). Impact of white wine macromolecules on fractionation performance of ultrafiltration membranes. *Journal International des Sciences de la Vigne et du Vin*, 45, 181-188.

Doco, T. Vuchot, P. Cheynier, V. and Moutounet, M. (2003). Structural modification of wine arabinogalactans during aging on lees. *American Journal of Enology and Viticulture*, 54, 150-157.

Dupin, I. V. S. McKinnon, B. M. Ryan, C. Boulay, M. Markides, A. J. Jones, G. P. Williams, P. J. and Waters, E. J. (2000). *Saccharomyces cerevisiae* mannoproteins that protect wine from protein haze: Their release during fermentation and lees contact and a proposal for their mechanism of action. *Journal of Agricultural and Food Chemistry*, 48, 3098-3105.

Esteruelas, M. Poinsaut, P. Sieczkowski, N. Manteau, S. Fort, M. F. Canals, J. M. and Zamora, F. (2009). Characterization of natural haze protein in sauvignon white wine. *Food Chemistry*, 113, 28-35.

Falconer, R. J. Marangon, M. Van Sluyter, S. C. Neilson, K. A. Chan, C. and Waters, E. J. (2010). Thermal stability of thaumatin-like protein, chitinase, and invertase isolated from Sauvignon blanc and Semillon juice and their role in haze formation in wine. *Journal of Agricultural and Food Chemistry*, *58*, 975-980.

Ferreira, R. B. Piçarra-Pereira, M. A. Monteiro, S. Loureiro, V. B. and Teixeira, A. R. (2002). The wine proteins. *Trends in Food Science and Technology*, *12*, 230-239.

Feuillat, M. (2003). Yeast macromolecules: Origin, composition, and enological interest. *American Journal of Enology and Viticulture*, *54*, 211-213.

Gonçalves, F. Heyraud, A. De Pinho, M. N. and Rimaudo, M. (2002). Characterization of white wine mannoproteins. *Journal of Agricultural and Food Chemistry*, *50*, 6097-6101.

Hsu, J. C. and Heatherbell, D. A. (1987). Heat-unstable proteins in wine. I. Characterization and removal by bentonite fining and heat treatment. *American Journal of Enology and Viticulture*, *38*, 11-16.

Jégou, S. Conreux, A. Villaume, S. Hovasse, A. Schaeffer, C. Cilindre, C. Van Dorsselaer, A. and Jeandet, P. (2009). One step purification of the grape vacuolar invertase. *Analytica Chimica Acta*, *638*, 75-78.

Jones, P. R. Gawel, R., Francis, I. L. and Waters, E. J. (2008). The influence of interactions between major white wine components on the aroma, flavor and texture of model white wine. *Food Quality Preference*, *19*, 596-607.

Laemmli, U. K. (1970). Cleavage of structural protein during the assembly of the head of Bacteriophage T4. *Nature*, *227*, 680-685.

Lambri, M. Dordoni, R. Giribaldi, M. Violetta, M. R. and Giuffrida, M. G. (2012). Heat-unstable protein removal by different bentonite labels in white wines. *LWT - Food Science and Technology*, *46*, 460-467.

Lobos, F. A. (2011). Characterization of Proteins Responsible for Protein Haze in Sauvignon Blanc Wines: Bioengineering Thesis. Concepción, Chile: University of Concepción.

Lomolino, G. and Curioni, A. (2007). Protein haze formation in white wines: Effect of *Saccharomyces cerevisiae* cell wall components prepared with different procedures. *Journal of Agricultural and Food Chemistry*, *55*, 8737-8744.

Lubbers, S. Voilley, A. Feuillat, M. and Charpentier, C. (1994). Influence of mannoproteins from yeast on the aroma intensity of a model wine. *LWT - Food Science and Technology*, *27*, 108-114.

Lubbers, S. Charpentier, C. and Feuillat, M. (1996). Etude de la rétention de composés d'arôme par les bentonites en mout, vin et milieux modèle. *Vitis*, *35*, 59-62.

Marchal, R. Bouquelet, S. and Maujean, A. (1996). Purification and partial biochemical characterization of glycoproteins in a champenois Chardonnay wine. *Journal of Agricultural and Food Chemistry, 44*, 1716-1722.

Marangon, M. Van Sluyter, S. C. Neilson, K. A. Chan, C. Haynes, P. Waters, E. J. and Falconer, R. J. (2011). Roles of grape thaumatin-like protein and chitinase in white wine haze formation. *Journal of Agricultural and Food Chemistry, 59*, 733-740.

Moine-Ledoux, V. and Dubourdieu, D. (1999). An invertase fragment responsible for improving the protein stability of dry white wines. *Journal of the Science of Food and Agriculture, 79*, 537-543.

Okuda, T. Fukui, M. Takayanagy, T. and Yokotsuka, K. (2006). Characterization of major stable proteins in Chardonnay wine. *Food Science and Technology Research, 12*, 131-136.

Pocock, K. F. Hayasaka, Y. McCarthy, M. G. and Waters, E. J. (2000). Thaumatin-like proteins and chitinases, the haze-forming proteins of wine, accumulate during ripening of grape (*Vitis vinifera*) berries and drought stress does not affect the final levels per berry at maturity. *Journal of Agricultural and Food Chemistry, 48*, 1637-1643.

Ribéreau-Gayon, P. Glories, Y. Maujean, A., and Dubourdieu, D. (2006). Clarification and stabilization treatments: fining wine. In P. Ribéreau-Gayon, Y. Glories, A. Maujean, and D. Dubourdieu (Eds.), Handbook of Enology. Volume 2. The Chemistry of Wine Stabilization and Treatments (2nd edition, pp. 301-332). Chichester: Wiley.

Rowe, J. D. Harbertson, J. F. Osborne, J. P. Freitag, M. Lim, J. and Bakalinsky, A.T. (2010). Systematic identification of yeast proteins extracted into model wine during aging on the yeast lees. *Journal of Agricultural and Food Chemistry, 58*, 2337-2346.

Sauvage, F. X. Bach, B. Moutounet, M. and Vernhet, A. (2010). Proteins in white wines: Thermo-sensitivity and differential adsorption by bentonite. *Food Chemistry, 118*, 26-34.

Vernhet, A. Pellerin, P. Belleville, M. P. Planque, J. and Moutounet, M. (1999). Relative impact of major wine polysaccharides on the performances of an organic microfiltration membrane. *American Journal of Enology and Viticulture, 50*, 51-56.

Vincenzi, S. Marangon, M. Tolin, S. and Curioni, A. (2011). Protein evolution during the early stages of white winemaking and its relations with wine stability. *Australian Journal of Grape and Wine Research, 17*, 20-27.

Waters, E. J. Wallace, W. Tate, M. E. and Williams, P. J. (1993). Isolation and partial characterization of a natural haze protective factor from wine. *Journal of Agricultural and Food Chemistry, 41*, 724-730.

Waters, E. J. Pellerin, P. and Brillouet, J. M. (1994). A Saccharomyces mannoprotein that protects wine from protein haze. *Carbohydrate Polymers, 23*, 185-191.

Waters, E. J. Alexander, G. Muhlack, R. Pocock, K. F. Colby, C. O'Neil, B. K. Høj, P. B. and Jones, P. (2005). Preventing protein haze in bottled white wine. *Australian Journal of Grape and Wine Research, 11*, 215-225.

Yokotsuka, K. Nozaki, K. and Takayanagy, T. (1994). Characterization of soluble glycoproteins in red wine. *American Journal of Enology and Viticulture, 45*, 410-416.

In: Macromolecular Chemistry: New Research   ISBN: 978-1-62417-854-2
Editor: Valentin Gartner   © 2013 Nova Science Publishers, Inc.

Chapter 4

PROTEIN TRANSDUCTION IN HUMAN CELLS MEDIATED BY ARGININE-RICH CELL-PENETRATING PEPTIDES IN MIXED COVALENT AND NONCOVALENT MANNERS

Betty Revon Liu[1], Ming-Huan Chan[2],
Hwei-Hsien Chen[3], Yue-Wern Huang[4]
and Han-Jung Lee[1,]*

[1]Department of Natural Resources and Environmental Studies,
National Dong Hwa University, Hualien, Taiwan
[2]Institute of Neuroscience, National Chengchi University,
Taipei City, Taiwan
[3]Institute of Population Health Sciences, National Health Research
Institutes, Miaoli, Taiwan
[4]Department of Biological Sciences,
Missouri University of Science and Technology, Rolla, US

ABSTRACT

Cell-penetrating peptides (CPPs) are small peptides with a high
content of basic amino acid residues. They possess the ability to

---

* Corresponding author: E-mail: hjlee@mail.ndhu.edu.tw.

translocate through the plasma membrane and facilitate exogenous cargo delivery into living cells. In this chapter, we demonstrate that arginine-rich CPPs are able to not only traverse cellular membranes by themselves, but also carry macromolecules into human A549 lung carcinoma cells in mixed covalent and noncovalent manners. This special macromolecular delivery system was named as mixed covalent and noncovalent protein transductions (CNPT). We found that cells treated with nona-arginine (R9)-red fluorescent protein (RFP) fusion protein mixed with green fluorescent protein (GFP), referred to as R9-RFP/GFP complexes, exhibit both red and green fluorescent images. Cells treated with R9-GFP fusion protein mixed with RFP, denoted as R9-GFP/RFP complexes, emitted green and red fluorescence, vice versa. Furthermore, mechanistic studies revealed that the cellular uptake mechanism of CNPT may involve a combination of multiple internalization pathways. Therefore, applications of this binary CNPT system may provide an efficient tool for delivery of multiple proteins in bioscience and clinical research.

ABBREVIATIONS

BFP   blue fluorescent protein
CNPT   covalent and noncovalent protein transductions
CPP   cell-penetrating peptide
CPT   covalent protein transduction
CytD   cytochalasin D
EIPA   5-(N-ethyl-N-isopropyl)-amiloride
GFP   green fluorescent protein
HR9   histidine-rich nona-arginine
NEM N-ethylmaleimide
NPT   noncovalent protein transduction
PBS   phosphate buffered saline
QD   quantum dot
R9   nona-arginine
RFP   red fluorescent protein
Tat   transactivator of transcription

INTRODUCTION

Cell-penetrating peptides (CPPs) were discovered from the transactivator of transcription (Tat) protein of the human immunodeficiency virus type 1

(HIV-1) to penetrate cells and activate viral genome replication [1, 2]. PTD, an 11-amino acid (amino acid sequence: YGRKKRRQRRR) sequence of Tat, was shown to be responsible for cellular internalization [3]. The process by which Tat protein and other CPPs cross the cell membrane and deliver macromolecule cargoes is referred to as protein transduction [4–6]. CPPs share a common character that is rich in basic amino acids or amphipathic, and can permeate membranes into living cells [4]. They are able to not only traverse cellular membranes by themselves, but also carry macromolecules, such as proteins, nucleic acids, liposomes, and nanoparticles, into cells [4].

We have demonstrated that nona-arginine (R9) peptides can deliver proteins into various types of cells in three manners: covalent protein transduction (CPT) [7–10], noncovalent protein transduction (NPT) [11–15], and combined association of covalent and noncovalent protein transductions (CNPT) synchronously [16, 17]. For CPT, CPPs can efficiently deliver covalently fused proteins, such as green fluorescent protein (GFP) and red fluorescent protein (RFP), into several types of cells. For NPT, CPPs are capable of efficiently delivering noncovalently associated proteins, RNAs, or DNAs into living cells or tissues in fully active forms. CPPs can deliver cargo proteins into living cells in CNPT.

Moreover, arginine-rich CPPs can also efficiently transport noncovalently associated biological macromolecules, such as DNAs [18–22] and RNAs [23], or quantum dots (QDs; nanoparticles) [24–26] into living cells or organisms. Recently, novel histidine-rich nona-arginine (HR9) peptides were demonstrated to deliver QDs at a higher efficiency than R9 peptides do [27]. Our studies indicated that the cellular entry mechanism for HR9/QD complexes was mediated by direct membrane translocation.

In this chapter, we demonstrate that arginine-rich R9 CPPs are able to deliver proteins into human cells by CNPT. Cellular internalization mechanism of the binary CNPT mediated by R9 is further investigated.

MATERIALS AND METHODS

Cell Culture

Human A549 lung carcinoma cells (American Type Culture Collection, Manassas, VA, USA; CCL-185) were maintained in Roswell Park Memorial Institute (RPMI) 1640 medium supplemented with 10% (v/v) fetal bovine serum (Gibco, Invitrogen, Carlsbad, CA, USA), as previously described [11].

Plasmid and Protein Preparation

pQE8-GFP, pR9-GFP, pR9-dTomato, and dTomato plasmids were described previously [11, 17].

In brief, the pQE8-GFP plasmid contains a *GFP* coding sequence under the control of the T5 promoter. Both pR9-GFP and pR9-dTomato plasmids consist of the *R9-GFP* and *R9-RFP* fusion-protein coding regions under the control of the T7 promoter, respectively.

The dTomato plasmid contains a *RFP* coding sequence under the control of the T7 promoter. Protein expression and purification from *Escherichia coli* were described previously [16, 28].

Covalent and Noncovalent Protein Transductions (CNPT)

For CPT, A549 cells were treated with phosphate buffered saline (PBS), 10 µM of GFP or dTomato as controls, or 30 µM of R9-GFP or R9-dTomato fusion protein at room temperature for 10 min and washed with PBS five times in order to remove free proteins.

For CNPT, cells were treated with the mixtures of 30 µM of R9-dTomato fusion protein and 10 µM of GFP, referred to as R9-RFP/GFP complexes, or 30 µM of R9-GFP fusion protein and 10 µM of dTomato, denoted as R9-GFP/RFP complexes, at a molar ratio of 3:1 at room temperature for 10 min and washed with PBS.

Flow Cytometry

Flow cytometric analysis was performed as previously described [14]. CNPT-treated cells were analyzed using a Cytomics FC500 flow cytometer (Beckman Coulter, Fullerton, CA, USA).

The FL1 filter (excitation 488 nm, emission 525 nm) was used for GFP detection, and the FL3 filter (excitation 488 nm, emission 615 nm) was used for RFP detection.

Results were then analyzed using CXP software (Beckman Coulter). C1 area represents only FL3 positive signals, C2 for both FL1 and FL3 positive signals, C3 for both FL1 and FL3 negative signals and C4 for only FL1 positive signals.

Fluorescent Microscopy

Fluorescent and bright-field images were observed and recorded using an Olympus IX70 inverted fluorescent microscope (Olympus, Center Valley, PA, USA), as previously described [17].

For the GFP detection, we set excitation at 460–490 nm and emission at 520 nm. For the RFP, we set excitation at 510–550 nm and emission at 590 nm. For the blue fluorescent protein (BFP), we set excitation at 330–385 nm and emission at 420 nm.

Images were captured by a Hamamatsu ORCA285 CCD camera. Shutters, filters, and camera were controlled using SlideBook software (Intelligent Imaging Innovations, Denver, CO, USA).

Mechanism of Cellular Uptake

Cells were treated with R9-GFP/RFP complexes at a molecular ratio of 3:1 in the absence or presence of physical treatment 4°C or pharmacological inhibitors 1 mM of N-ethylmaleimide (NEM), 100 μM of 5-(N-ethyl-N-isopropyl)-amiloride (EIPA), 10 μM of cytochalasin D (CytD), and 80 mM of sodium chlorate ($NaClO_3$) (Sigma-Aldrich, St. Louis, MO, USA), as previously described [16, 24].

Nucleic acids of cells were then stained with Hoechst 33342 (Invitrogen, Carlsbad, CA, USA), as previously described [24]. GFP, RFP, and BFP channels indicated the distribution of R9-GFP, dTomato, and the location of nucleus, respectively.

Fluorescent and bright-field images were recorded using a fluorescent microscope.

Statistical Analysis

Results are expressed as mean ± standard deviation. Mean values and standard deviations were calculated from at least three independent experiments carried out in triplicates per treatment group. Statistical comparisons between the control and treated groups were performed by the Student's t-test, using levels of statistical significance of $P < 0.05$ (*, †) and $P < 0.01$ (**, ††), as indicated.

RESULTS

To examine CNPT in cells, A549 cells treated with R9-RFP/GFP complexes were analyzed by flow cytometry. Fluorescence could not be detected in the cells treated with either PBS (Figure 1A) or GFP alone (Figure 1B).

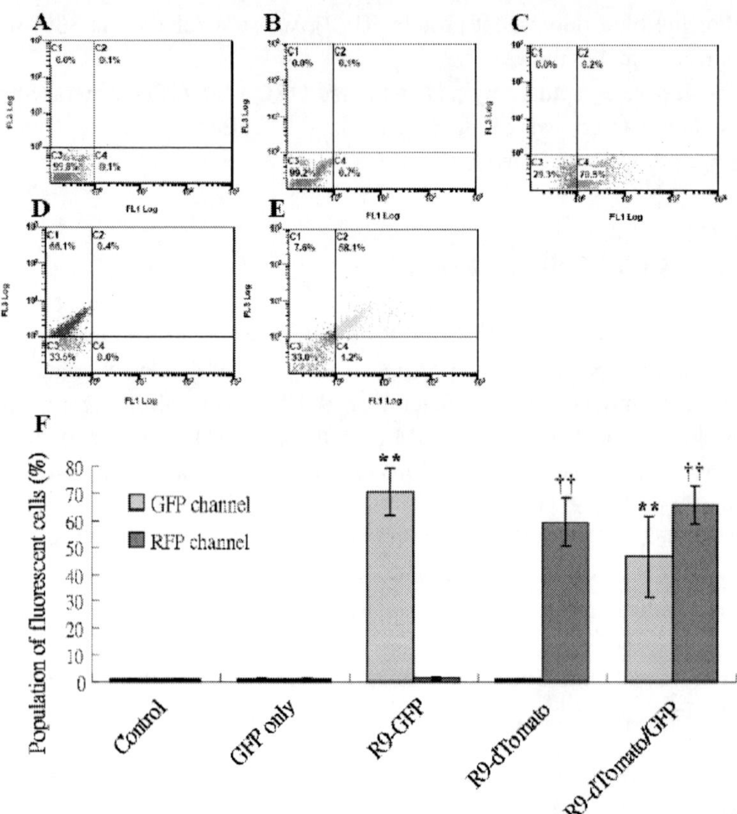

Figure 1. Flow cytometric analysis of CNPT in A549 cells. Cells were treated with PBS as a negative control (A), 10 μM of GFP (B), 30 μM of R9-GFP (C), 30 μM of R9-dTomato (D), or the mixtures of 30 μM of R9-dTomato fusion protein and 10 μM of GFP, referred to as R9-RFP/GFP complexes at a molar ratio of 3:1 (E). FL1 filter was used for GFP detection, and FL3 filter was used for RFP detection. C1 area represents only FL3 positive signals, C2 for both FL1 and FL3 positive signals, C3 for both FL1 and FL3 negative signals and C4 for only FL1 positive signals. (F) Histogram of fluorescent cells in populations. Significant differences of $P < 0.05$ (* for the GFP channel or † for the RFP channel) and $P < 0.01$ (** or ††) are indicated.

Cells treated with R9-GFP, serving as a positive control of CPT, exhibited 70.5% in green fluorescence (Figure 1C and 1F). Cells treated with R9-RFP showed 66.1% in red fluorescence (Figure 1D and 1F). In contrast, cells treated with R9-RFP/GFP complexes displayed 58.1% double positive signals (in C2 area) in both green and red fluorescence (Figure 1E and 1F). These results demonstrate that R9 peptides not only transport covalently fused proteins into cells by themselves, but also deliver noncovalently associated proteins into living cells.

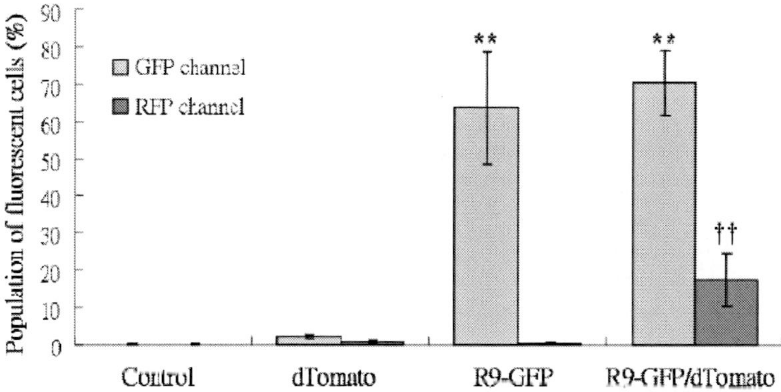

Figure 2. Histogram of flow cytometric analysis of CNPT. A549 cells were treated with PBS (control), 10 µM of dTomato, 30 µM of R9-GFP, or 30 µM of R9-GFP fusion protein and 10 µM of dTomato, denoted as R9-GFP/RFP complexes, at a molar ratio of 3:1. The populations of fluorescent cells were analyzed by a flow cytometer. Significant differences of $P < 0.05$ (* for the GFP channel or † for the RFP channel) and $P < 0.01$ (** or ††) are indicated.

To extend the CNPT mediated by R9 peptides to other cargo proteins, we switched fluorescent protein partners of carrier and cargo. Little signal was detected in A549 cells treated with either PBS (control) or RFP alone (Figure 2). Cells treated with R9-GFP significantly displayed green fluorescence. However, cells treated with R9-GFP/RFP complexes for 10 min exhibited both green and red fluorescent images. These results confirm again the internalization ability of R9 peptides to deliver different proteins into cells in CNPT.

To investigate potential mechanism of R9 peptides in CNPT, cells were treated with R9-GFP/RFP complexes in the presence or absence of 4°C treatment, clathrin- and caveolae-mediated uptake inhibitor NEM, macropinocytosis inhibitor EIPA, F-actin polymerization inhibitor CytD, and proteoglycan sulfation inhibitor sodium chlorate.

In the 4°C treatment, all energy-dependent molecular movements are believed to be inhibited [29]. We found that the treatment of low temperature incubation, NEM, and $NaClO_3$ reduced protein internalization in NPT, but did not alter cellular uptake in CPT (Figure 3).

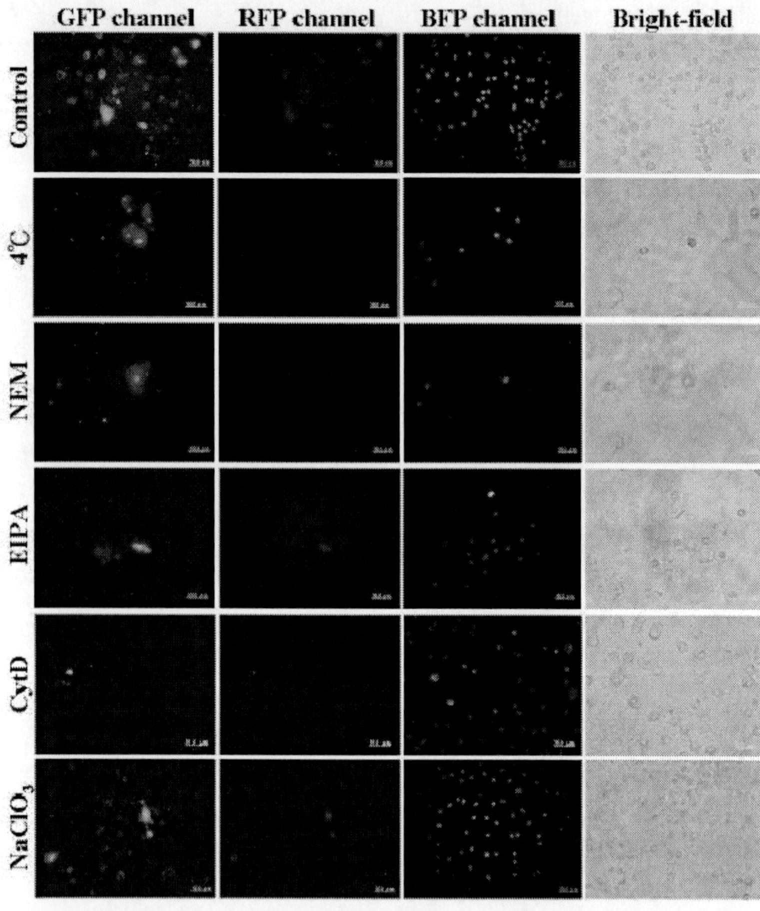

Figure 3. Investigation of mechanism of CNPT. A549 cells were treated with R9-GFP/RFP complexes at a molecular ratio of 3:1 in the absence or presence of inhibitors (4°C, NEM, EIPA, CytD, and sodium chlorate). Cells were then stained with Hoechst 33342. GFP, RFP, and BFP channels indicate the distribution of R9-GFP, dTomato, and the location of nucleus, respectively. Fluorescent and bright-field images were observed and recorded using a fluorescent microscope. Scale bar is 50 μm.

These results indicate that CPT does not require energy for cellular uptake and include the possibility of direct membrane translocation mechanism in CPT. On the contrary, NPT requires energy for cellular uptake and includes the possibility of an endocytic mechanism in NPT.

Treatments of macropinocytosis inhibitor EIPA and cytoskeleton modulator CytD inhibited cellular uptake in both CPT and NPT (Figure 3). These data indicate that CNPT may involve macropinocytosis. Taken together, these studies reveal that the cellular uptake mechanism of binary CNPT may involve a complicated combination of multiple internalization pathways.

DISCUSSION

In this chapter, we demonstrate that R9 peptides not only transport covalently fused proteins into cells, but also simultaneously deliver noncovalently associated proteins into cells. Mechanistic studies revealed that the cellular uptake mechanism of CNPT may involve a combination of multiple internalization pathways.

Despite the lack of clear understanding of the cellular uptake mechanism, the current model for CPP internalization is suggested to be a multistep process [29]. The process of protein transduction involves at least the interaction between CPP/cargo and the cell surface, the translocation of CPP/cargo through the cell membrane, and finally the target distribution of cargo into the cytoplasm, organelle, or even nucleus. Recently, it was generally believed that most CPPs utilize two or multiple pathways for cellular entry [30–32]. The two major uptake mechanisms of CPPs are direct membrane translocation and endocytosis. Direct membrane translocation, also named direct penetration, is a fast and energy-independent pathway. Endocytosis is an energy-dependent pathway and consists of phagocytosis and pinocytosis.

We found in the present study that CNPT may involve two major uptake mechanisms, direct membrane translocation and endocytosis, with the treatment of 4°C, NEM, and $NaClO_3$ in cells (Figure 3). The uptake of CPPs in human HeLa cells was suppressed by the macropinocytosis inhibitor EIPA and the F-actin polymerization inhibitor CytD, which suggests a role for macropinocytosis in CPP uptake [33]. In the present study, R9 peptide-mediated CNPT was inhibited by the treatment of CytD and EIPA (Figure 3). These results demonstrate that CNPT is dependent on macropinocytosis and actin rearrangement. In summary, our results of mechanistic studies are consistent with recent reviews [34–37] indicating potential uptake mechanism of protein cargoes mediated by R9 peptides in CNPT involves a complicated combination of multiple internalization pathways.

CONCLUSION

In conclusion, we demonstrate that arginine-rich R9 peptides are able to deliver cargo proteins into living cells in CNPT. The cellular internalization mechanism of CNPT may involve a complicated combination of multiple internalization pathways. These results of binary CNPT provide a useful strategy to synchronously deliver two or more biological macromolecules in drug delivery.

ACKNOWLEDGMENTS

We thank Dr. Roger Y. Tsien (University of California, San Diego, CA, USA) for provision of dTomato plasmid, and Dr. Michael B. Elowitz (California Institute of Technology, CA, USA) for pQE8-GFP plasmid. This work was supported by the Postdoctoral Fellowship NSC 101-2811-B-259-001 (to B.R.L.) and the Grant Number NSC 101-2320-B-259-002-MY3 (to H.-J.L.) from the National Science Council, Taiwan.

REFERENCES

[1]   Green, M. and Loewenstein, P. M. (1988). Autonomous functional domains of chemically synthesized human immunodeficiency virus tat trans-activator protein. *Cell*, *55*, 1179–1188.

[2]   Frankel, A. D. and Pabo, C. O. (1988). Cellular uptake of the tat protein from human immunodeficiency virus. *Cell*, *55*, 1189–1193.

[3]   Vives, E., Brodin, P. and Lebleu, B. (1997). A truncated HIV-1 Tat protein basic domain rapidly translocates through the plasma membrane and accumulates in the cell nucleus. *J. Biol. Chem.*, *272*, 16010–16017.

[4]   Wadia, J. S. and Dowdy, S. F. (2002). Protein transduction technology. *Curr. Opin. Biotechnol.*, *13*, 52–56.

[5]   Deshayes, S., Konate, K., Aldrian, G., Crombez, L., Heitz, F. and Divita, G. (2010). Structural polymorphism of non-covalent peptide-based delivery systems: highway to cellular uptake. *Biochim. Biophys. Acta*, *1798*, 2304–2314.

[6]   Gautam, A., Singh, H., Tyagi, A., Chaudhary, K., Kumar, R., Kapoor, P. and Raghava, G. P. (2012). CPPsite: a curated database of cell penetrating peptides. *Database*, 2012, bas015.

[7]   Chang, M., Chou, J. C. and Lee, H. J. (2005). Cellular internalization of fluorescent proteins via arginine-rich intracellular delivery peptide in plant cells. *Plant Cell Physiol.*, *46*, 482–488.

[8]   Liu, K., Lee, H. J., Leong, S. S., Liu, C. L. and Chou, C. J. (2007). A bacterial indole-3-acetyl-L-aspartic acid hydrolase inhibits mung bean (*Vigna radiata* L.) seed germination through arginine-rich intracellular delivery. *J. Plant Growth Regul.*, *26*, 278–284.

[9]   Li, J. F., Huang, Y., Chen, R. L. and Lee, H. J. (2010). Induction of apoptosis by gene transfer of human *TRAIL* mediated by arginine-rich intracellular delivery peptides. *Anticancer Res.*, *30*, 2193–2202.

[10]   Liou, J. S., Liu, B. R., Martin, A. L., Huang, Y. W., Chiang, H. J. and Lee, H. J. (2012). Protein transduction in human cells is enhanced by cell-penetrating peptides fused with an endosomolytic HA2 sequence. *Peptides*, *37*, 273–284.

[11]   Wang, Y. H., Chen, C. P., Chan, M. H., Chang, M., Hou, Y. W., Chen, H. H, Hsu, H. R. and Lee, H. J. (2006). Arginine-rich intracellular delivery peptides noncovalently transport protein into living cells. *Biochem. Biophys. Res. Commun.*, *346*, 758–767.

[12]   Chang, M., Chou, J. C., Chen, C. P., Liu, B. R. and Lee, H. J. (2007). Noncovalent protein transduction in plant cells by macropinocytosis. *New Phytol.*, *174*, 46–56.

[13]   Hou, Y. W., Chan, M. H., Hsu, H. R., Liu, B. R., Chen, C. P., Chen, H. H. and Lee, H. J. (2007). Transdermal delivery of proteins mediated by non-covalently associated arginine-rich intracellular delivery peptides. *Exp. Dermatol.*, *16*, 999–1006.

[14]   Liu, B. R., Chou, J. C. and Lee, H. J. (2008). Cell membrane diversity in noncovalent protein transduction. *J. Membr. Biol.*, *222*, 1–15.

[15]   Liu, B. R., Huang, Y. W., Chiang, H. J. and Lee, H. J. (2012). Primary effectors in the mechanisms of transmembrane delivery of arginine-rich cell-penetrating peptides. *Adv. Stu. Biol.*, in press.

[16]   Hu, J. W., Liu, B. R., Wu, C. Y., Lu, S. W. and Lee, H. J. (2009). Protein transport in human cells mediated by covalently and

noncovalently conjugated arginine-rich intracellular delivery peptides. *Peptides*, *30*, 1669–1678.

[17] Lu, S. W., Hu, J. W., Liu, B. R., Lee, C. Y., Li, J. F., Chou, J. C. and Lee, H. J. (2010). Arginine-rich intracellular delivery peptides synchronously deliver covalently and noncovalently linked proteins into plant cells. *J. Agricult. Food Chem.*, *58*, 2288–2294.

[18] Chen, C. P., Chou, J. C., Liu, B. R., Chang, M. and Lee, H. J. (2007). Transfection and expression of plasmid DNA in plant cells by an arginine-rich intracellular delivery peptide without protoplast preparation. *FEBS Lett.*, *581*, 1891–1897.

[19] Dai, Y. H., Liu, B. R., Chiang, H.J. and Lee, H. J. (2011). Gene transport and expression by arginine-rich cell-penetrating peptides in *Paramecium*. *Gene*, *489*, 89–97.

[20] Lee, C. Y., Li, J. F., Liou, J. S., Charng, Y. C., Huang, Y. W. and Lee, H. J. (2011). A gene delivery system for human cells mediated by both a cell-penetrating peptide and a *piggyBac* transposase. *Biomaterials*, *32*, 6264–6276.

[21] Chen, Y. J., Liu, B. R., Dai, Y. H., Lee, C. Y., Chan, M. H., Chen, H. H. and Lee, H. J. (2012). A gene delivery system for insect cells mediated by arginine-rich cell-penetrating peptides. *Gene*, *493*, 201–210.

[22] Liu, B. R., Lin, M. D., Chiang, H. J. and Lee, H. J. (2012). Arginine-rich cell-penetrating peptides deliver gene into living human cells. *Gene*, *505*, 37–45.

[23] Wang, Y. H., Hou, Y. W. and Lee, H. J. (2007). An intracellular delivery method for siRNA by an arginine-rich peptide. *J. Biochem. Biophys. Methods*, *70*, 579–586.

[24] Liu, B. R., Li, J. F., Lu, S. W., Lee, H. J., Huang, Y. W., Shannon, K. B. and Aronstam, R. S. (2010). Cellular internalization of quantum dots noncovalently conjugated with arginine-rich cell-penetrating peptides. *J. Nanosci. Nanotechnol.*, *10*, 6534–6543.

[25] Liu, B. R., Huang, Y. W., Chiang, H. J. and Lee, H. J. (2010). Cell-penetrating peptide-functionized quantum dots for intracellular delivery. *J. Nanosci. Nanotechnol.*, *10*, 7897–7905.

[26] Xu, Y., Liu, B. R., Chiang, H. J., Lee, H. J., Shannon, K. B., Winiarz, J. G., Wang, T. C., Chiang, H. J. and Huang, Y. W. (2010). Nona-arginine facilitates delivery of quantum dots into cells via multiple pathways. *J. Biomed. Biotechnol.*, *2010*, 948543.

[27] Liu, B. R., Huang, Y. W., Winiarz, J. G., Chiang, H. J. and Lee, H. J. (2011). Intracellular delivery of quantum dots mediated by a histidine-

and arginine-rich HR9 cell-penetrating peptide through the direct membrane translocation mechanism. *Biomaterials, 32,* 3520–3537.

[28] Chang, M., Hsu, H. Y. and Lee, H. J. (2005). Dye-free protein molecular weight markers. *Electrophoresis, 26,* 3062–3068.

[29] Kaplan, I. M., Wadia, J. S. and Dowdy, S. F. (2005). Cationic TAT peptide transduction domain enters cells by macropinocytosis. *J. Contr. Release, 102,* 247–253.

[30] Nakase, I., Kobayashi, S. and Futaki, S. (2010). Endosome-disruptive peptides for improving cytosolic delivery of bioactive macromolecules. *Biopolymers, 94,* 763–770.

[31] Schmidt, N., Mishra, A., Lai, G. H. and Wong, G.C. (2010). Arginine-rich cell-penetrating peptides. *FEBS Lett., 584,* 1806–1813.

[32] Madani, F., Lindberg, S., Langel, U., Futaki, S. and Graslund, A. (2011). Mechanisms of cellular uptake of cell-penetrating peptides. *J. Biophys., 2011,* 414729.

[33] Nakase, I., Niwa, M., Takeuchi, T., Sonomura, K., Kawabata, N., Koike, Y., Takehashi, M., Tanaka, S., Ueda, K., Simpson, J. C., Jones, A. T., Sugiura, Y. and Futaki, S. (2004). Cellular uptake of arginine-rich peptides: roles for macropinocytosis and actin rearrangement. *Mol. Ther., 10,* 1011–1022.

[34] Chung, S. K., Maiti, K. K. and Lee, W. S. (2008). Recent advances in cell-penetrating, non-peptide molecular carriers. *Int. J. Pharm., 354,* 16–22.

[35] Hansen, M., Kilk, K. and Langel, U. (2008). Predicting cell-penetrating peptides. *Adv. Drug Deliv. Rev., 60,* 572–579.

[36] Nakase, I., Takeuchi, T., Tanaka, G. and Futaki, S. (2008). Methodological and cellular aspects that govern the internalization mechanisms of arginine-rich cell-penetrating peptides. *Adv. Drug Deliv. Rev., 60,* 598–607.

[37] Koren, E. and Torchilin, V. P. (2012). Cell-penetrating peptides: breaking through to the other side. *Trends Mol. Med., 18,* 385–393.

In: Macromolecular Chemistry: New Research   ISBN: 978-1-62417-854-2
Editor: Valentin Gartner   © 2013 Nova Science Publishers, Inc.

Chapter 5

CLEAVAGE OF FIBRINOPEPTIDES FROM FIBRINOGEN DURING FIBRIN FORMATION

Eduard Brynda and Tomas Riedel

Institute of Macromolecular Chemistry
Academy of Sciences of the Czech Republic
v.v.i., Prague, the Czech Republic

ABSTRACT

Fibrinogen, one of the most abundant proteins in the blood, plays a key role in haemostasis, inflammation, wound healing, and some other physiological and pathological processes.

HAEMOSTASIS

Immediately after a damage of blood vessel endothelium the tissue factor pathway of coagulation starts by bringing blood into contact with subendothelial structures. A cascade of molecular interactions among coagulation factors dissolved in blood plasma is initiated by the formation of a complex between factor VII and tissue factor (TF) located on membranes of cells within and around the vessel wall. The VIIa/TF complex activates factors IX and X and the proceeding process results with the production of thrombin from its zymogen prothrombin circulating in blood. TF is absent on the surface of undisturbed vascular endothelial cells composing the inner lining of blood vessels and its concentration increases from smooth muscle cells in the media to fibroblasts in the adventitia [1-3]. Simultaneously with the TF pathway, platelets, which do not adhere to the normal endothelial cells, adhere to subendothelium due to interactions between their surface membrane receptors (integrins) and components of the exposed extracellular matrix (collagen, fibronectin and laminin) and via interaction with vonWillebrand factor (vWF) adsorbed on the subendothelium from blood plasma. Once adherent to subendothelium, platelets spread out on the surface, and additional platelets, delivered by the flowing blood, adhere first to the basal layer of adherent platelets and eventually to one another. A crucial event in platelet aggregation is induction of a change in the disposition of membrane receptor GP IIb/IIIa [4], which acquires the capacity to bind fibrinogen from blood plasma. Fibrinogen mediates platelet aggregation by forming bridges from platelet to platelet via these receptors. Whereas vWF and collagen can interact with resting platelets, fibrinogen forms a high-affinity bond only with GP IIb/IIIa on activated platelets [5]. TF pathway generates only a small amount of thrombin because it is rapidly shut down by TF pathway inhibitor and antithrombin [6-8] The initially formed traces of thrombin are not detectable using a standard chromogenic substrate in model experiments, but they are sufficient to activate platelets and factors XI, VIII, and V, which induce the consolidation pathway rapidly generating a mass of thrombin [9-11]. The consolidation pathway is mediated by membrane receptors exposed on the surface of activated platelets that are specific the for binding plasma coagulation factors FXI [12,13], FXIa [14], FIX [15], FIXa [15,16], FVIII [17], FVIIIa [17], FX [17], FXa [19], FV[20], FVa [20], prothrombin [21], and thrombin [22]. These receptors promote the assembly of coagulation enzyme–cofactor–substrate complexes that, compared with solution-phase reactions, vastly accelerate the activation of factor XI by factor XIIa [23] or thrombin [9, 12] and the activation of factor X by factor IXa [24,25]. thereby providing an efficient catalytic environment for the conversion of prothrombin to thrombin by factor Xa [26-28]. In addition, cofactors for prothrombin activation, factor Va and negatively charged phospholipids, from interior of normal platelets appear on surface of activated platelets. The generated thrombin catalyzes conversion of fibrinogen brought by circulating blood to fibrin. Fibrin monomers associate to polymeric fibrin structures that reinforce the primary plug of aggregated platelets and form a fibrin clot in the plug vicinity. Factor XIII activated by platelet-generated thrombin crosslinks fibrin polymers by covalent bonds between c-chains and Aa-chains of their fibrin monomer units. These cross-links stabilize fibrin contacts within the fibrin clot and increase the elasticity of individual fibrin fibers present within the clot [29]. The clot is farther protected against fibrinolysis by thrombin-activated fibrinolysis inhibitor (TAFI). Antithrombin III (ATIII) helps to localize coagulation at the site of injury by inhibiting free thrombin and some other coagulation factors, such as, as, activated forms of FX (Xa), FIX (IXa), FXI (XIa), FXII (XIIa), and FVII, released into circulating blood [30-33] during the thrombus formation.

Later the fibrin clot provides a temporary scaffold for adhesion and stimulation of cells that take part in the regeneration of damaged vessel, such as, fibroblasts, smooth muscle cells within the vessel wall, and endothelial cells creating a new endothelium. [34]. Cells adhere to fibrin directly via interaction of their membrane receptors, mainly integrin avb3, with tripeptide amino acid sequences, e.g. arginine-glycine-aspartic acid (RGD), present in fibrin structure. The indirect cell adhesion is based on the capacity of fibrin to attach fibronectin and vitronectin from circulating blood. [35]. The cellular response is further regulated by a number of bioactive molecules accumulated by fibrin, such as, basic fibroblast growth factor and vascular endothelial growth factor [36.], or enzymes and proenzymes, e.g., plasminogen activator and plasminogen [37]. Ultimately, fibrin clot is degraded by proteolytic action of plasmin and matrix metalloproteinases and gradually replaced by extracellular matrix produced by the adhered cells [38].

COAGULATION AT FOREIGN SURFACES

Contrary to the tissue factor pathway of coagulation, contact activation (intrinsic) pathway is initiated by interaction of blood plasma with a foreign surface. The contact activation complex is believed to involve four proteins; coagulation factor XII (FXII, Hageman factor), prekallikrein (PK, Fletcher factor), high-molecular weight kininogen (HMWK, Fitzgerald factor), and coagulation factor XI (FXI, Plasma Thromboplastin Antecedent) [39]. As this pathway does not need participation of platelets, it is responsible for in vitro formation of fibrin in plasma free of platelets., The undesirable growth of thrombi on surfaces of all known biomaterials used in blood-contacting medical devices [28,29] is probably due to both the initiation of the contact pathway and platelet-mediated reactions, in proportions that presumably depend on the surface chemistry of the biomaterial and characteristics of the blood flow in which the biomaterial is immersed [26,27]. According to a widely accepted hypothesis the process starts with kallikrein (Kal)-mediated activation of FXII adsorbed to negatively charged surfaces [40]. On the other hand, plasma coagulates less quickly also on hydrophilic surfaces that are not negatively charged as well as on hydrophobic surfaces. New findings indicate that surface-catalyzed conversion of the blood zymogen factor XII to the enzyme FXIIa is not specific for anionic/hydrophilic surfaces, as proposed by the traditional theory [41-43]. Even if initiation of this pathway by plasma contact with extracellular matrix below the damaged endothelium cannot be excluded, the main role of the TF pathway in the formation of haemostatic plug at a site of injury in vivo is illustrated by the fact that patients with severe deficiencies of FXII, HMWK, and prekallikrein do not have a bleeding disorder.

FIBRINOGEN TO FIBRIN CONVERSION

The conversion of soluble fibrinogen to fibrin starts as soon as thrombin is produced by coagulation processes in blood or blood plasma or if it is added to any fibrinogen containing medium in vitro. Structural properties of fibrinogen and thrombin play a key role in this process.

Fibrinogen is a 340-kDa plasma glycoprotein with a complex structure. Fibrinogen molecule consists of two identical subunits each composed of three non-identical polypeptide chains, Aα, Bβ and γ [44]. These chains are linked together by 29 disulfide bonds and form several structural regions, two distal D regions, one central E region, and two αC regions [45]. X-ray studies of fibrinogen crystals [46-48] revealed that in the central region the disulfide-linked NH2-terminal portions of all six chains form central nodule while in each distal D region the COOH-terminal portions of the Bβ and γ chains form globular β- and γ-nodules, respectively. Each pair of the distal nodules is linked with the central nodule by a triple helical coiled-coil connector composed of the middle portions of all three chains. The COOH-terminal portions of the Aα chains (αC regions) fold back to the central E region to form two interacting αC-domains [35]. The central nodule contains two pairs of polymerization sites (knobs), "A" and "B", the complementary polymerization sites (holes) "a" and "b" are located in the distal γ- and β-nodules, respectively. Besides, these nodules as well as the αC-domains contain numerous binding sites that become active after conversion of fibrinogen into fibrin or adsorption of fibrinogen on various surfaces [50-52].

Mechanisms of fibrinogen - fibrin conversion and subsequent formation of fibrin network have been studied primarily by experiments in which thrombin has been added to fibrinogen solutions in buffers. It is postulated that thrombin binds to the central region of fibrinogen and catalyzes cleavage of two short peptides, the fibrinopeptide A (FpA) and the fibrinopeptide B (FpB), located at the NH2-termini of the Aα and Bβ chains, respectively [53]. Using chromatography five variants of the fibrinopeptides (A, AP, Y, B and B-R) were isolated from human fibrinogen. Fibrinopeptide A (FpA) consists of 16 amino acid residues. Its sequence using single letter nomenclature is as follows: ADSGEGDFLAEGGGVR. Molecular weight of FpA is 1536Da. Fibrinopeptide AP is an FpA phosphorylated at the serine residue in position 3 from the N-terminal end. Fibrinopeptide Y is an FpA without first amino acid residue from the N-terminal end. FpAP yields approximately 1/3 of that of FpA. FpY constitute the smallest component of the fibrinopeptides. Fibrinopeptide B (FpB) consists of 14 amino acid residues and its sequence using single letter nomenclature is as follows: PyrGGVNDNEEGFFSAR. It has been found, that the first amino acid residue of the FpB is pyroglutamic acid (PyrG). Molecular weight of FpB is 1552Da. Fibrinopeptide B-R is an FpB without the last amino acid Arg that is cleaved off by carboxypeptidase B. The level of FpB-R may account up to 1/3 of fibrinopeptide B. [54] High resolution spectra of FpA complex with thrombin have been identified by NMR and X-ray crystal analysis. NMR spectra of fibrinopeptide A with thrombin shows that residues 7-16 form a compact structure interacting with thrombin. On the other hand, first five amino acid residues do not interact with thrombin. [55,56] Binding of FpA into thrombin pocket is facilitated by a short helical segment of the FpA, a chain reversal at Ala at position 10 and multiple hydrogen bonds within residues 7-12. Furthermore, Phe at position 8, Leu at position 9 and Val at position 15 form a hydrophobic group that binds to the thrombin apolar pocket. The side chain of Arg at position 16 interacts with thrombin pocket (P1). Fibrinopeptide A sequences within Gly at position 6 to Arg at position 16 is conserved in many species, suggesting that this region is critical for interaction with thrombin.

The importance of residues Asp at position 7 and Phe at position 8 of the FpA was shown previously. Moreover, thrombin does not cleave the FpA when Phe at position 8 is replaced by Tyr in fibrinopeptide A. [57,58]The most critical amino acid is Arg at position 16, which is the position where thrombin cleaves off the FpA. Replacement of this amino acid leads to a slower or completely impaired FpA release catalyzed by thrombin. Arg at position 16 is one of the most frequently replaced amino acid residue in fibrinogen. Up to date 89 mutations were found in the position 16 replacing Arg with Cys, His or Ser. Their clinical outcome varies from asymptomatic to hemorrhages or even to thromboses. However, only one mutation was found at position 17 exchanging Gly to Val. Such replacement had only mild impact on the fibrinopeptide release. FpA were released slower than control and the patient displayed mild bleeding disorders. [59] Binding of FpB to the fibrinogen is facilitated by Phe at the position 10 and 11 that binds to the apolar thrombin pocket in the opposite direction to FpA. According to the NDSK-derived peptide Bbeta 1-118 FpB seems to be poor substrate compared with FpA. Up to date 14 different mutations were found in the position 14 and 15 replacing Arg and Gly with Cys. It is quite surprising that most of the found mutations were connected with thrombotic complications. [59]

Fibrinopeptide A and B often serve as an assay providing an index of thrombin action on fibrinogen. These assays are used to help diagnose severe problems with blood clotting. In general, the level of FpA should range from 0.6 to 1.9 ng/ml, however normal values may slightly differ among different laboratories. Increased FpA levels may be connected with diseases such as disseminated intravascular coagulation (DIC), certain types of leukemia, some infections, cellulitis and system lupus erythematosis. Elevated levels of FpB are often connected with pulmonary embolism, deep vein thrombosis and other thrombotic events.

The cleavage of FpA and FpB exposes knobs "A" and "B", which interact with complementary holes "a" and "b" of neighboring molecules to form fibrin polymer [60]. According to the current view, the assembly of fibrin clot in solution occurs in two steps. First, thrombin removes FpA enabling "A"-"a" interactions between individual half-staggered molecules and resulting in formation of two-stranded protofibrils. Next, protofibrils aggregate laterally to make thicker fibers that coalesce and branch to form a three-dimensional fibrin network. It was shown that lateral aggregation of protofibrils coincides with the removal of FpB enabling "B"-"b" interactions and that FpB removal is accelerated by forming polymers [61,62]. Thus, during fibrin polymerization in solution the release of FpA occurs very rapidly while the release of FpB is delayed and reaches its maximum when fibrin formation is almost completed. Such a delay results in a sequential release of fibrinopeptides and thereby sequential activation of the two sets of polymerisation sites [63,64], which is necessary for normal fibrin assembly [62]. On the other hand, FpB release is not essential for lateral aggregation of protofibris. Fibrin clot formation can also occur following the release of only FpA by several snake venom enzymes. The resulting des-FpA fibrin molecules assemble into protofibrils and subsequently undergo lateral aggregation despite the lack of FpB cleavage and B:b interactions[65-67]. Binding of thrombin to the fibrinogen E region, which occurs through its anion-binding exosite I [68-70], plays a major role in substrate recognition and removal of FpA and FpB; it may also be involved in their sequential cleavage. Numerous studies performed to characterize this binding culminated in solution of the crystal structure of thrombin in complex with the isolated E region of fibrin [71], which revealed structural details of the interaction between these molecules. As to the mechanism of sequential cleavage of fibrinopeptides, it was proposed that higher thrombin specificity for FpA accounts for the delay in FpB release, a lag phase, during fibrin polymerization and that conformational changes accompanying FpA release facilitate FpB cleavage [72]. Further, structural analysis of the thrombin-E complex [71] revealed that in fibrinogen the FpA-containing portions of the A□ chains are in a more favorable position to bind in the active site cleft of bound thrombin thus providing the structural basis for the preferential cleavage of FpA at the first step of fibrin assembly [73]. It was also suggested that the accelerating effect of polymer formation on FpB cleavage is related mainly to the interaction of N-terminal portions of the Bβ chains with the dimeric DD regions formed in protofibrils and fibers which brings FpB into the vicinity of bound thrombin [73].

FORMATION OF POLYMER FIBRIN STRUCTURES

The formation of polymer fibrin structures has been usually studied in by measuring turbidity in fibrinogen-thrombin solutions. No turbidity increase was detected during initial formation of protofibrils in a ''lag'' phase at the beginning of the fibrinogen conversion to fibrin. Lack of homogeneity in population of fibrin structures was indicated by light scattering [74,75]. Transmission electron micrographs of dried samples showed fibrin monomers and small double-stranded structures, such as, dimers, trimers, tetramers, and protofibrils made up of 8 or more monomers At the middle of the lag phase, the fraction of protofibrils became elevated compared with the fraction of dimers, trimers, and fibers [76]. This observation suggests that during this time the rate of longitudinal growth of oligomers and protofibrils is higher than the rate of lateral aggregation of these structures [77]. Subsequent "lateral" side-by-side aggregation of protofibrils causes an increase in turbidity. To proceed from longitudinal to lateral growth, protofibrils must be sufficiently long, approximately in the range of 0.6-0.8 μm.[78] Next, the fibrin network is established through fiber branching and lateral and longitudinal growth, resulting in a gel [79-81. Finally, there is some network rearrangement due to the formation of new oligomers and fibers after the gel point [79,82]. The rate of polymerization is affected by many factors, including the rate of fibrinopeptide cleavage, oligomer formation, and lateral aggregation [83]. The mass/length ratio of the fibrin fibers present in clot was determined by turbidity measurements owing to the length of fibers greater than the wavelength of visible light.[84]. The structure was also observed using scanning or transmission electron microscopy or confocal microscopy [85,86.].

The formation of fibrin clot in vivo depends on the thrombin production. A "lag" phase before appearing an observable fibrin clot formation includes (i) reactions at the surface of TF-bearing cells confined to the damaged vessel wall producing an amount of thrombin, which is believed to be incapable of generating fibrin clot, (ii) the transfer of thrombin to platelets adhered to the damaged vessel lumen, and (iii) initiation of the coagulation cascade at the platelet surface. During the next "propagation" phase the amount of produced thrombin sharply increases due to its generation at the surface of activated platelets. A massive thrombin generation leads to the observable formation of fibrin clot.

The research on relationships between various types of haemostatic disorders and coagulation factor deficiency in patients' blood and model experiments with the factor-deficient plasma makes it possible to reveal the role of molecular factors participating in haemostasis. On the other hand, fibrin formation during blood coagulation at a site of injury cannot be observed in real time at molecular level. Thus, current knowledge has been mainly based on observing processes, such as, fibrinogen-to-fibrin conversion, protofibril formation, and their lateral aggregation, induced by adding thrombin into fibrinogen solutions.

In our recent studies we have focused on some disregarded processes that might take part in the formation of fibrin clot.

ACTIVITY OF THROMBIN BOUND TO FIBRIN OR IMMOBILIZED FIBRINOGEN

Mechanisms regulating the extension of fibrin away from surfaces of activated platelets or artificial biomaterials are still not well understood. The inhibition of thrombin and other coagulation factors by ATIII diffusing to the vicinity of the active surfaces, the opposite diffusion of generated thrombin to the circulating blood [87], and a high affinity binding of free thrombin to the formed fibrin gel [88,89] are supposed to retard the distant clot formation. Thrombin (m.w. 34,000) has a catalytic protease centre and two positively charged patches located at the opposite ends of the molecule called anion-binding exosites I and II.[90] Exosite I (fibrinogen recognition site) binds fibrinogen and thrombin inhibitor hirudin. Exosite II binds heparin which accelerates the inhibition of thrombin by ATIII. During fibrin gel formation, thrombin can be bound to fibrin in two orientations. With the first orientation thrombin is bound via both the fibrin recognizing exosites and the catalytic centre. Such binding decreases concentration of active thrombin in the growing clot. With the second orientation thrombin is bound only via the exosites while the catalytic centre remains accessible to fibrinogen. Thrombin bound in the latter orientation remains catalytically active. Unlike free thrombin, this bound thrombin is protected against inhibition by ATIII, however, it can be inactivated by ATIII-independent inhibitors, such as, hirudin and D-phenylalanyl-L-prolyl-L-arginine-chloromethylketone (PPACK).[91-93].

Surface plasmon resonance (SPR) together with turbidity measurements allowed us to observe in situ processes associated with activity of thrombin bound to fibrinogen or fibrin gel immobilized on solid substrates [94]. Adsorption of Fbg on gold surface from Fbg solutions in Tris buffer (TB) flowing along the surface was observed in real time by SPR. After washing the Fbg coated surface with TB and phosphate buffered saline (PBS) the adsorbed Fbg was treated with thrombin solution in PBS. The SPR measurement was not sensitive enough to detect thrombin binding to the surface, however, the activity of thrombin that remained attached to the surface after washing it with PBS was detectable using chromogenic substrate. A fast increase in a mass due to fibrin formation at the surface was observed immediately after the thrombin treated Fbg layer was contacted with a fresh Fbg solution. The mass kept increasing until Fbg solution was replaced with TB and fibrin gel remained attached at the surface. Evidently, the fibrin formation was catalyzed by thrombin bound to the a fibrinogen/fibrin layer immobilized on the surface. While SPR measurements monitored the processes within the penetration depth of evanescent field at the SPR chip surface (about 100 nm), the formation of thicker fibrin gels could be observed using turbidity measurement. The same procedure as it was used in SPR flow cell was performed in a quartz cuvette. In these experiments the inner surface of a quartz cuvette was coated with adsorbed Fbg, treated with thrombin, and washed with buffer. In the first step of the experiment shown in figure 1 an increase in turbidity started, when the thrombin treated Fbg surface was contacted with Fbg solution.

In the second step the Fbg solution was replaced with the buffer. The turbidity slightly decreased and then persisted constant indicating that a fibrin clot remained on the cuvette surface. In the third step the buffer was replaced with fresh Fbg solution and the turbidity continued increasing (Figure 1, solid curve). A portion of Fbg solution was taken from the cuvette during the first step and transferred into a clean cuvette. The increase in turbidity continued in this solution indicating that fibrin structures were formed, even if no thrombin was added (Figure 1, dashed curve). An increase in turbidity was observed also if Fbg solution containing ATIII (Fbg/ATIII) was contacted with the cuvette surface previously coated by fibrinogen and treated with thrombin (Figure 2, curve 2).

No fibrin formation was observed in the Fbg/ATIII solution transferred into a clean cuvette. ATIII did not inhibit thrombin bound to the fibrinogen/fibrin layer immobilized on the surface, however, it could inhibit thrombin released form the surface into the ambient solution. A gradual release of the bound thrombin into the ambient Fbg solution probably slowed down and finally stopped the growth of the fibrin gel at the surface visible in all curves in figure 2.

The thrombin release was confirmed independently by a gradual decrease in catalytic activity of the surface with the bound thrombin measured by chromegenic substrate. As expected, no formation of the surface fibrin gel was observed if bound thrombin was inhibited by PPACK and hirudin. The structure of fibrin gel produced by catalytic action of thrombin bound to adsorbed fibrinogen/fibrin layer on ambient Fbg solution containing AT III was observed by atomic force microscopy (AFM) . A mash of tiny fibrin fibers is visible below the thick fibrin strands in figure 3. Only the tiny fibrin mash was observed if Fbg solution containing AT III and heparin (Fbg/ATIII/heparin) was used for growth of the surface fibrin gel.

Probably the tiny structure was generated directly by the bound thrombin while the thicker strands were formed by thrombin released into Fbg solution that was not inhibited by ATIII sufficiently fast but completely inhibited in the presence of ATIII and heparin.

The growth of the surface fibrin gel from heparinized blood plasma is shown in figure 2, curve 3. Surface of the cuvette was initially coated by incubation with heparinized blood plasma prepared by adding 2.5 mM CaCl2 and heparin, 60 µg/mL, into citrated human blood plasma. After incubation with thrombin in TB solution, the surface was washed with TB and PBS and contacted with the heparinized plasma again. The gel formation was observed by the increase in turbidity.

No increased turbidity was observed in the plasma transferred after the gel formation into a clean cuvette. Thus, the gel was formed only on the thrombin treated surface. The curve 3 in figure 2 reflecting the growth of surface fibrin gel in the heparinized blood plasma is quite similar to the curve 4 reflecting the same procedure in which solution of Fbg, ATIII, and heparin in TB was used..

Probably, fibrinogen present in a protein layer adsorbed from blood plasma was responsible for binding thrombin capable to convert fibrinogen in the ambient plasma to fibrin. It could be supposed that such process leads to the formation of some fibrin gel also when thrombin is produced in vivo at the blood contact with synthetic surfaces.

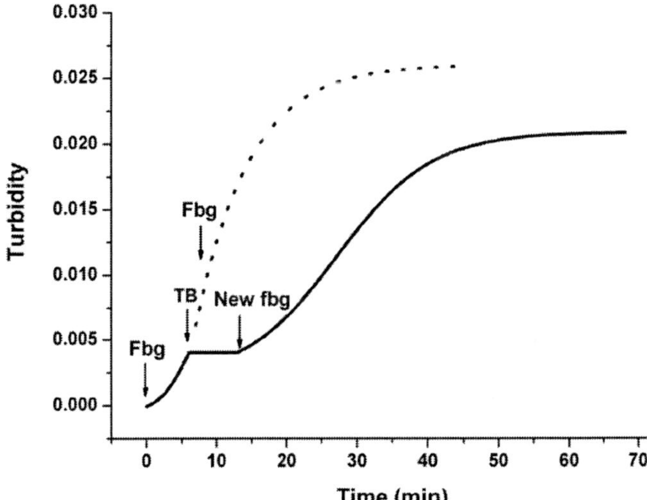

Figure 1. The formation of fibrin gel by catalytic action of thrombin attached to a fibrinogen/fibrin layer adsorbed on surface of a a quartz cuvette. Surface was initially coated by adsorption from fibrinogen solution (2 µg/mL for 90 min), incubated with thrombin solution (2.5 NIH U/mL for 10 min), and washed with buffer. Solid line: Fbg - Turbidity was measured after buffer was replaced with fibrinogen solution (200 µg /mL) TB - Fibrinogen solution was replaced with buffer. New fbg - Buffer was replaced with a fresh fibrinogen solution. Dotted line: A sample of fibrinogen solution taken from the coated cuvette was transferred into a clean cuvette where turbidity was measured.

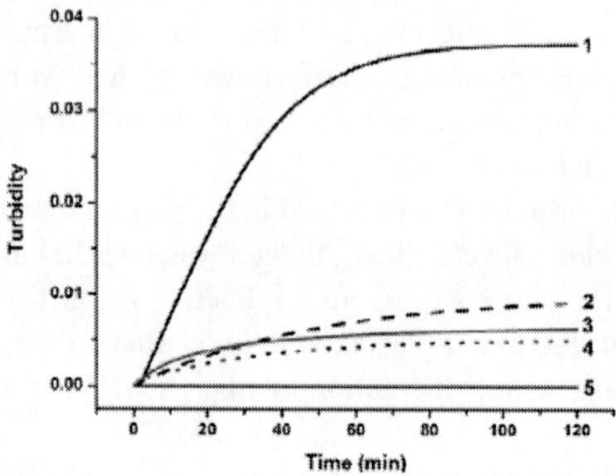

Figure 2. The formation of fibrin gel by catalytic action of thrombin attached to surface of a quartz cuvette initially coated by adsorption from fibrinogen solution or citrated human blood plasma. Surface was initially coated by adsorption from fibrinogen solution (2 µg/mL for 90 min) for experiments 1,2,4,and 5 or citrated blood plasma for curve 3, washed with buffer, incubated with thrombin solution (2.5 NIH U/mL for 10 min), and washed with buffer; turbidity was measured after the buffer was replaced with a solution containing fibrinogen or blood plasma Curve 1—fibrinogen solution, 200 µg/mL; Curve 2—fibrinogen solution 200 µg/mL containing AT III 0.5 NIU/mL; Curve 3—citrated human blood plasma supplemented with 2.5 mM CaCl2 and heparin 60 µg/mL; Curve 4—fibrinogen solution 200 µg/mL containing AT III 0.5 NIU/mL, and heparin 60 µg/mL; Curve 5—surface with attached thrombin was treated with hirudin, 6 NIH U/mL, and PPACK 10 µg/mL for 15 was introduced.

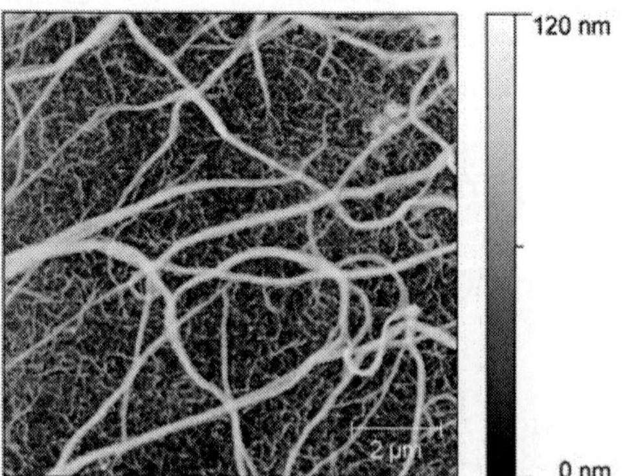

Figure 3. AFM micrograph of freeze-dried surface fibrin gel produced by catalytic action of thrombin bound to adsorbed fibrinogen/fibrin layer on ambient fibrinogen solution containing AT III. Fibrinogen was initially adsorbed from concentration 2 µg/mL for 60 min on a glass surface and treated with 2.5 NIH U/mL thrombin for 10 min. Then the surface was incubated with solution of fibrinogen 200 µg/mL containing AT III 0.5 U/mL for 120 min.

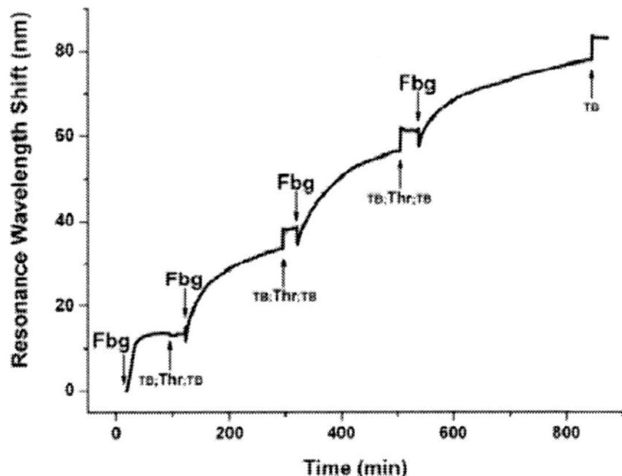

Figure 4. The successive growth of surface fibrin gel by alternating incubation the surface with thrombin solution and fibrinogen solution containing ATIII and heparin measured by SPR. The surface was initially coated by adsorption from fibrinogen solution, 2 µg/mL (Fbg), incubated with thrombin solution (Thr) and. washed with buffer (TB), then the procedure cinsisting in the incubation with fibrinogen solution containing AT III 0.5 NIU/mL and heparin 10 IU/ml,(Fbg), washing with buffer, and incubation with thrombin solution was successively repeted. Arrows indicate the replacement of solutions.

In the experiment shown in figure 4, a surface fibrin gel was prepared from Fbg/ATIII/heparin solution, washed with buffers, and treated with thrombin. A renewed growth of fibrin gel started when the surface was subsequently contacted with a fresh Fbg/ATIII/heparin solution. By repeating the procedure a step growth of fibrin gel was observed. When thrombin is generated in vivo by some of the coagulation processes, the generation of fibrin by catalytic action of thrombin attached to fibrin gel or to immobilized fibrinogen might accompany the standard fibrin formation catalyzed by free thrombin or prolong the process after free thrombin was inhibited by ATIII. The later process could be effective particularly at the distant margin of fibrin clot to which ATIII is readily delivered by blood circuit.

IMMOBILIZED FIBRINOGEN AND FIBRIN-FIBRINOGEN INTERACTIONS

Except of several specific non-fouling surfaces fibrinogen adsorbs on any artificial surface [95,96]. Fibrinogen was found in protein layers adsorbed on hydrophobic or hydrophilic surfaces by contact with blood plasma [97]. It binds also to platelets via integrin GP IIb/IIIa. Adsorption of fibrinogen at the site of vascular injury and inflammation, or on artificial surfaces of vascular grafts and other blood-contacting components of medical devices plays a significant role in its interaction with cells and in blood coagulation [98,99]. Numerous studies have demonstrated that fibrinogen in solution and fibrinogen adsorbed on various surfaces exhibit different properties [100-103]. For example, surface-adsorbed fibrinogen changes its conformation and thus reveals multiple binding sites that interact with the receptors on platelets and leukocytes [104,105]. These reciprocal interactions participate in the process of blood clot formation and in inflammatory response. Platelet adhesion promoted by the deposition of fibrinogen might contribute to the development of the inflammatory response during ischemia reperfusion. Fibrinogen binding to the intercellular adhesion molecule 1 (ICAM-1), expressed on endothelial cells, has been reported to mediate the adhesion of leukocytes to these cells in vitro, and to induce the attachment of monocytic cells to the vascular wall of mesenteric venules in vivo. The structural properties of fibrinogen play a key role in its interactions with various biomolecules and cell types. Using infrared multi-internal reflection spectroscopy and monoclonal antibodies against fibrinogen E and D regions, we have previously found that fibrinogen adsorbs on glass, carbon, polyethylene, and polystyrene surfaces in basically two different orientation: "side-on" (laying on the surface) and "end-on" (standing on the surface) [106-108]. While side-on orientation prevails during adsorption from solutions with low fibrinogen concentrations; end-on orientation, in which adsorbed fibrinogen is closely-packed, occurs with high fibrinogen concentrations in solution.

In our recent experiments [109] Fbg adsorption at low and high concentration, conversion of adsorbed Fbg to fibrin by thrombin from ambient solution, and the formation of dimer and trimer structures by the attachment Fbg to immobilized fibrin were investigated using Attenuated Total Reflection Fourier Transform Infrared Spectroscopy (ATR FT-IR), SPR, and by measuring kinetics of the release of fibrinopeptides FpA and FpB during the processes. The amount of Fbg adsorbed on polystyrene surface at low Fbg concentration of 20 µg/mL in solution determined by ATR FT-IR was about a half of that adsorbed at a high concentration of 500 µg /mL indicating the preferential side-on and end-on orientation of adsorbed Fbg molecules, respectively. After washing the surface with TB and PBS the coated surface was incubated with thrombin solution in PBS. The release of fibrinopeptides FpA and FpB cleaved by thrombin from adsorbed Fbg was detected by HPLC analysis of samples taken from the thrombin solution at selected times of incubation. When the primary Fbg layer was formed by Fbg adsorption from the hig concentration (presumably "end-on" adsorption), the release of both fibrinopeptides occurred very rapidly and stopped after 60 min. The initial rates of release of both fibrinopeptides were similar; however, the amount of released FpA was more than 40% lower than that of FpB (17 versus 30 pmol). In contrast, when the primary layer was formed by the adsorption of fibrinogen from the lower concentration (presumably "side-on" adsorption), the initial rate of FpA release was higher than that of FpB. Nevertheless, in 60 min the amount of released FpB approached that of FpA (about 10 pmol). The kinetics resembled FpA and FpB release from fibrinogen in solution; except of the delay, a lag phase, in FpB release characteristic for the latter. Contrary to the closely packed Fbg molecules adsorbed end-on to the surface, central parts E of Fbg molecules adsorbed side-on the surface were exposed to thrombin (Figure 5a) similarly to those of Fbg molecules in solution (Figure 5b). It was proposed that higher thrombin specificity for FpA accounts for the delay in FpB release during fibrin polymerization in solution and that conformational changes accompanying FpA release facilitate FpB cleavage [50, 110]. Such changes seem to occur already due to Fbg adsorption. It was also suggested that the accelerating effect of polymer formation on FpB cleavage in solution is related mainly to the interaction of N-terminal portions of the Bβ chains with the dimeric DD regions formed in protofibrils and fibers, which brings FpB to the vicinity of bound thrombin [73]. The presence of analogous dimeric DD regions between neighbor Fbg adsorbed side-on the surface (Figure 5a) and/or the conformation changes of Fbg molecules induced by adsorption could explain the absence of the lag phase in the FpB release from adsorbed Fbg. If we inhibited surface-bound thrombin remaining on the surface after the release of FpA and FpB from the primary adsorbed Fbg layer by PPACK and hirudin, an additional attachment of Fbg to the surface was observed by SPR when the surface was incubated with Fbg solution. The binding could have been via interaction of holes "a" and "b" located in the D regions of the added fibrinogen, and the complementary knobs "A" and "B" of the E regions created by cleaving FpA and FpB from the initially adsorbed Fbg (Figure 5a). When the surface with secondary immobilized Fbg. was incubated with thrombin, the release of both fibrinopeptides occurred rapidly and reached a maximum after 60 min, with the amount of released FpA being significantly lower than that of FpB (Figure 6). At the same time, no delay of FpB release (lag-phase) was observed. After the bound thrombin was inhibited, the surface was again incubated with a fresh Fbg solution and the additional binding of Fbg to the new fibrin formed in the secondary layer of immobilized Fbg was observed (Figure 5a). Short fibrous structures were observed by transmission electron microscopy on the surface prepared by this procedure on thin carbon film.

The sequential fibrin formation based on fibrinogen to fibrin binding connected with the preferential release of FpB might occur also during in vivo formation of fibrin clot, particularly, in the distant regions where fibrin is generated by thrombin in excess of fibrinogen and ATIII delivered by circulating blood.

The formation of fibrin dimers, trimers, and protofibrils via binding and subsequent conversion of Fbg (Figure 5b) could be an alternative way to the standard in vivo generation of such fibrin structures that further asssociate to fibrin gel. The conversion of Fbg bound to fibrin strands (Figure 5a) in the fibrin gel could contribute to the clot growth.

It was shown that the overall structure of fibrin arising with the preferential release of FpB by a specific snake venom enzyme is essentially the same as the structure of fibrin initiated by thrombin [111]. Also no delay in the release of FpB was observed by Blombäck et all. [112,113], In their experiments with whole blood and platelet-rich plasma, in which FpB was released almost as quickly as FpA.

To explain these observations, they suggested that the binding of fibrinogen to the platelet receptor GPIIb/IIIa may induce conformational changes in bound fibrinogen, resulting in the exposure of a thrombin-susceptible cleavage site and thus facilitating FpB release.

A physiological relevance of fibrinogen binding to the outer region of fibrin clot was indicated earlier by a reduced adhesiveness of fibrinogen-fibrin matrices towards leukocytes and platelets [49,50]. The reduced adhesiveness was attributed to the altered physical properties of the immobilized fibrinogen, which creates an anti-adhesive fibrinogen shell, limiting thrombus growth and protecting the thrombus from leukocyte attachment and from premature dissolution by leukocyte proteases.

As FpB is a potent chemoattractant [42]; its preferential release from the fibrinogen attached to fibrin clot may indicate the physiological purpose in the attraction of cells to the site of injury. The thrombin-mediated exposure of neo-epitopes at the N-terminus endows the formed fibrin, as opposed to fibrinogen, with the ability to bind to endothelial cell surface receptors, and thereby induce the endothelium regeneration [114,115]

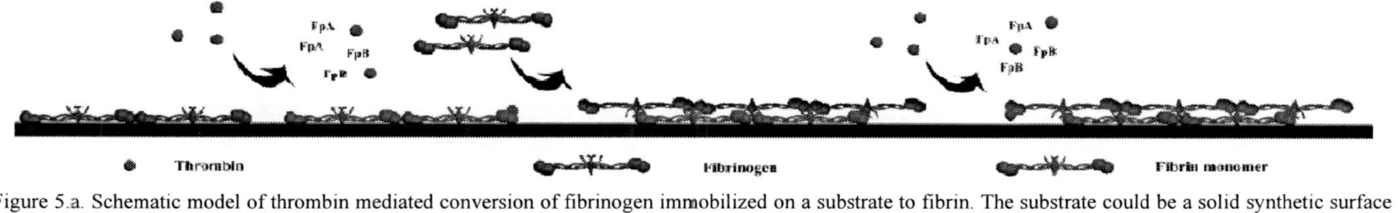

Figure 5.a. Schematic model of thrombin mediated conversion of fibrinogen immobilized on a substrate to fibrin. The substrate could be a solid synthetic surface or strands in a fibrin gel.

Figure 5b. The formation of fibrin dimers and ptotofibrils in fibrinogen solutions based on binding fibrinogen to fibrin followed with conversion of the bound fibrinogen to fibrin - an alternative way to standard protofibril formation via association of fibrin monomers..

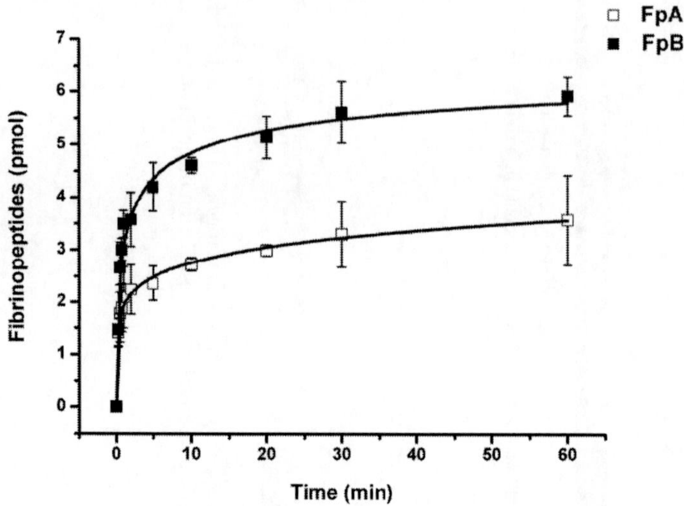

Figure 6. Kinetics of thrombin-mediated release of fibrinopeptides from fibrinogen attached to fibrin. Fibrinogen-fibrin complexes were formed by the immobilization of the secondary fibrinogen layer on thrombin-treated (2.5 U/ml thrombin for 60 min, followed by a mixture of PPACK and hirudin for 20 min) primary fibrinogen layers adsorbed onto polystyrene. The primary layer was adsorbed from 20 µg/ml fibrinogen for 150 min; the secondary layer was immobilized at 200 µg/ml fibrinogen for 60 min. (The amounts of FpA (open squares) and FpB (solid squares) released from the secondary layers upon incubation with 2.5 U/ml thrombin were determined by HPLC. Each point represents a mean value obtained from three independent experiments.

Our observations describe above revealed some mechanisms that could take place in the formation of fibrin clot at a site of vessel wall injury or at contact of blood with surfaces of synthetic materials. (i) Fibrin formation by cleavage of fibrinopeptides from fibrinogen present in a protein layer adsorbed from blood plasma on a substrate followed by the growth of a fibrin network by catalytic action of thrombin bound to such immobilized fibrin could provide an alternative to the standard formation of fibrin clot when blood contacts a foreign surface. (ii) The growth of a fibrin network by catalytic action of thrombin bound to fibrin gel could accompany the standard clot formation by spontaneous association of fibrin structures formed in plasma. (iii) The catalytic action of thrombin bound to fibrin clot, which is not inhibited by ATIII, could be ceased by the gradual thrombin release. (iv) The attachment of fibrinogen to fibrin monomers, dimmers, protofibrils, and fibrin gel followed by the conversion of the attached fibrinogen to fibrin could provide an alternative to the standard formation of the fibrin structures from fibrin monomers produced in plasma.

REFERENCES

[1] Osterud B, Tindall A, Brox JH, Olsen JO. Thromboplastin content in the vessel walls of different arteries and organs of rabbits. *Thromb. Res.* 1986;42:323–9.

[2] Drake TA,Morrissey JH, Edgington TS. Selective cellular expression of tissue factor in human tissues. Implications for disorders of hemostasis and thrombosis. *Am. J. Pathol.* 1989;134:1087–97.

[3] Flossel C, Luther T, Muller M, Albrecht S, Kasper M. Immunohistochemical detection of tissue factor (TF) on paraffin sections of routinely fixed human tissue. *Histochemistry.* 1994;101:449–53.

[4] Shattil SJ, Kashiwagi H, Pmpori N. Integrin signaling the platelet paradigm., *Blood.* 1998, 91, 2645-2657.

[5] Bennett JS, Vilaire G, Exposure of platelet fibrinogen receptors by ADP and epinephire, *J. Clin. Invest.* 1979, 64, 1393-1401.

[6] Broze GJJ, Miletich JP. Characterization of the inhibition of tissue factor in serum. *Blood.* 1987;69:150–155.

[7] Rao LVM, Rapaport SI. Studies on the mechanisms of inactivation of the extrinsic pathway of coagulation. *Blood.* 1987;69:645–651.

[8] Sanders NL, Bajaj SP, Zivelin A, et al. Inhibition of tissue factor/factor VIIa activity in plasma requires factor X and an additional plasma component. *Blood.* 1985;66:204–212.

[9] Baglia FA, Walsh PN. Thrombin-mediated feedback activation of factor XI on the activated platelet surface is preferred over contact activation by factor XIIa or factor XIa. *J. Biol. Chem.* 2000;275:20514–20519.

[10] Gailani D, Broze GJ Jr. Factor XI activation in a revised model of blood coagulation. *Science.* 1991;253:909–912.

[11] Baglia FA, Badellino KO, Li CQ, et al. Factor XI binding to the platelet glycoprotein Ib-IX-V complex promotes factor XI activation by thrombin. *J. Biol. Chem.* 2002;277:1662–1668.

[12] Baglia FA, Shrimpton CN, Lopez JA, et al. The glycoprotein Ib-IX-V complex mediates localization of factor XI to lipid rafts on the platelet membrane. *J. Biol. Chem.* 2003;278:21744–21750.

[13] Sun MF, Baglia FA, Ho D, et al. Defective binding of factor XI-N248 to activated human platelets. *Blood.* 2001;98:125–129.

[14] Ahmad SS, London FS, Walsh PN. Binding studies of the enzyme (factor IXa) with the cofactor (factor VIII) in the assembly of factor-X activating complex on the activated platelet surface. *J. Thromb. Haemost.* 2003;1: 2348–2355.

[15] London FS, Walsh PN. Zymogen factor IX potentiates factor IXa-catalyzed factor X activation. *Biochemistry.* 2000;39:9850–9858.

[16] Wilkinson FH, London FS, Walsh PN. Residues 88-109 of factor IXa are important for assembly of the factor-X activating complex. *J. Biol. Chem.* 2002;277:5725–5733.

[17] Ahmad SS, London FS, Walsh PN. Binding studies of the enzyme (factor IXa) with the cofactor (factor VIII) in the assembly of factor-X activating complex on the activated platelet surface. *J. Thromb. Haemost.* 2003;1: 2348–2355.

[18] Ahmad SS, Walsh PN. Coordinate binding studies of the substrate (factor X) with the cofactor (factor VIII) in the assembly of the factor X activating complex on the activated platelet surface. *Biochemistry.* 2002;41: 11269–11276.

[19] Tracy PB, Nesheim ME, Mann KG. Platelet factor Xa receptor. *Methods Enzymol.* 1992;215:329–360.

[20] Tracy PB, Eide LL, Mann KG. Human prothrombinase complex assembly and function on isolated peripheral blood cell populations. *J. Biol. Chem.* 1985;260:2119–2124.

[21] Scandura JM, Ahmad SS, Walsh PN. A binding site expressed on the surface of activated human platelets is shared by factor X and prothrombin. *Biochemistry.* 1996;35:8890–8902.

[22] Ahmad SS, London FS, Walsh PN. The assembly of the factor X-activating complex on activated human platelets. *J. Thromb. Haemost.* 2003;1: 48–59.

[23] Walsh PN, Griffin JH. Contributions of human platelets to the proteolytic activation of blood coagulation factors XII and XI. *Blood.* 1981;57: 106–118.

[24] Baglia FA, Walsh PN. Thrombin-mediated feedback activation of factor XI on the activated platelet surface is preferred over contact activation by factor XIIa or factor XIa. *J. Biol. Chem.* 2000;275:20514–20519.

[25] Rawala-Sheikh R, Ahmad SS, Monroe DM, et al. Role of gammacarboxyglutamic acid residues in the binding of factor IXa to platelets and in factor-X activation. *Blood.* 1992;79:398–405.

[26] London FS, Walsh PN. Zymogen factor IX potentiates factor IXa-catalyzed factor X activation. *Biochemistry.* 2000;39:9850–9858.

[27] Hoffman M, Monroe DM, Oliver JA, Roberts HR. Factors IXa and Xa play distinct roles in tissue factor-dependent initiation of coagulation. *Blood.* 1995;86:1794–801.

[28] Bouchard BA, Catcher CS, Thrash BR, et al. Effector cell protease receptor1, a platelet activation-dependent membrane protein, regulates prothrombinase-catalyzed thrombin generation. *J. Biol. Chem.* 1997;272: 9244–9251.

[29] Liu W, Jawerth LM, Sparks EA, et al. Fibrin fibers have extraordinary extensibillity and elasticity. *Science.* 2006;313:634.

[30] Pike RN, Buckle AM, le Bonniec BF, Church FC. Control of the coagulation system by serpins. Getting by with a little help from glycosaminoglycans. *FEBS J.* 2005;272:4842–51.

[31] Quinsey NS, Greedy AL, Bottomley SP, Whisstock JC, Pike RN. Antithrombin: in control of coagulation. *Int. J. Biochem. Cell Biol.* 2004;36:386–9.

[32] Olds RJ, Lane DA, Mille B, Chowdhury V, Thein SL. Antithrombin: the principal inhibitor of thrombin. *Semin. Thromb. Hemost.*1994;20:353–72.

[33] Rao LV, Nordfang O, Hoang AD, Pendurthi UR. Mechanism of antithrombin III inhibition of factor VIIa/tissue factor activity on cell surfaces. Comparison with tissue factor pathway inhibitor/factor Xa-induced inhibition of factor VIIa/tissue factor activity. *Blood.* 1995;85:121–9.

[34] Laurens N, Koolwijk P, De Maat PM. Fibrin structure and wound healing. *J. Thromb. Haemost.* 2006;4:932–939.

[35] Makogonenko E, Tsurupa G, Ingham K, Medved L. Interaction of fibrin (ogen) with fibronectin: Further characterization and localization of the fibronectin-binding site. *Biochemistry.* 2002;41:7907–7913.

[36] Sahni A, Francis CW. Vascular endothelial growth factor binds to fibrinogen and fibrin and stimulates endothelial cell proliferation. *Blood.* 2000;96:3772–3778.

[37] Tsurupa G, Medved L. Fibrinogen C domains contain cryptic plasminogen and tPA binding sites. *Ann. N.Y. Acad. Sci.* 2001;936: 328–330.

[38] Clark RAF. Fibrin and wound healing. *Ann. N.Y. Acad. Sci.* 2001;936:355–367.

[39] Williams DF. Blood physiology and biochemistry: Hemostasis and thrombosis. In: Blood Compatibility. Williams DF. Eds. CRC Press. Boca Raton, Florida. 1987;5–35.

[40] Griep MA, Fujikawa K, Nelsestuen GL. Possible basis for the apparent surface selectivity of the contact activation of human blood coagulation factor XII. *Biochemistry*. 1986;25:6688–94.

[41] E. A. Vogler, C. A. Siedlecki. Contact activation of blood-plasma coagulation. *Biomaterials*. 2009;30:1857–69.

[42] Zhuo R, Siedlecki CA, Vogler EA. Autoactivation of blood factor XII at hydrophilic and hydrophobic surfaces. *Biomaterials*. 2006;27:4325–32.

[43] Chatterjee K, Guo Z, Vogler EA, Siedlecki CA. Contributions of contact activation pathways of coagulation factor XII in plasma. *J. Biomed. Mater Res. A.* 2009;90(1):27-34.

[44] Henschen AH, McDonagh J. Fibrinogen, fibrin and factor XIII. In: Blood Coagulation. Zwaal RFA and Hemker HC. Eds. Elsevier, Amsterdam. 1986;171–241.

[45] Medved L, Weisel JW. Recommendations for nomenclature on fibrinogen and fibrin. *J. Thromb. Haemost.* 2009;7:355-9.

[46] Brown JH, Volkmann N, Jun G, Henschen-Edman AH, Cohen C. The crystal structure of modified bovine fibrinogen. *P. Natl. Acad. Sci. U.S.A.* 2000;97:85-90.

[47] Yang Z, Kollman JM, Pandi L, Doolittle RF. Crystal structure of native chicken fibrinogen at 2.7 angstrom resolution. *Biochemistry*. 2001;40:12515-23.

[48] Kollman JM, Pandi L, Sawaya MR, Riley M, Doolittle RF. Crystal structure of human fibrinogen. *Biochemistry*. 2009;48:3977-86.

[49] Weisel JW, Medved L. The structure and function of the alpha C domains of fibrinogen. *Ann. N.Y. Acad. Sci.* 2001;936:312-27.

[50] Hu WJ, Eaton JW, Ugarova TP, Tang L. Molecular basis of biomaterial-mediated foreign body reaction. *Blood.* 2001; 98,1231-8.

[51] Lishko VK, Kudryk B, Yakubenko VP, Yee VC, Ugarova TP. Regulated unmasking of the cryptic binding site for integrin alpha(M)beta(2) in the gamma C-domain of fibrinogen. *Biochemistry*. 2002;41:12942-51.

[52] Medved L, Nieuwenhuizen W. Molecular mechanisms of initiation of fibrinolysis by fibrin. *Thromb. Haemostasis*. 2003;89:409-19.

[53] Blomback B. Fibrinogen and fibrin-proteins with complex roles in hemostasis and thrombosis. *Thromb. Res.* 1996; 83:1-75.

[54] Blombäck B, Blombäck M, Edman P, Hessel B. Human fibrinopeptides isolation, characterization and structure. *Biochimica et biophysica acta.* 1966:115;371-396.

[55] Ni F, Konishi Y, Frazier RB, Scheraga HA, Lord ST: High resolution NMR studies of fibrinogen-like peptides in solution: Interaction of

thrombin with residues 1-23 of the Aa chain of human fibrinogen. *Biochemistry.* 1989, 28:3082.

[56] Martin PD, Robertson W, Turk D, Huber R, Bode W, Edwards BF: The structure of residues 7-16 ofthe Aa-chain of human fibrinogen bound to bovine thrombin at 2.3-A resolution. *J. Biol. Chem.* 1992:267;7911.

[57] Marsh H.C. Jr., Meinwald Y.C., Lee S., Scheraga H.A. Biochemistry. 1983:22; 4170-4174.; Henschen A., Lottspeich F., Kahl M., Southan C. *Ann. N.Y. Acad. Sci.* 1983:408;28-43.

[58] Rooney MM, Mullin JL, Lord ST. Substitution of thyrosine for phenylalanine in fibrinopeptide A results in preferential thrombin cleavage of fibrinopeptide B from fibrinogen. *Biochemistry.* 1998:37;13704-13709.

[59] Hanss M, Biot F. A database for human fibrinogen variants. *Ann. N.Y. Acad. Sci.* 2001:936;89-90. Release. 36 01/02/2011.

[60] Yang Z, Mochalkin I, Doolittle RF. A model of fibrin formation based on crystal structures of fibrinogen and fibrin fragments complexed with synthetic peptides. *P. Natl. Acad. Sci. U.S.A.* 2000;97:14156-61.

[61] Weisel JW. Fibrin assembly – lateral aggregation and the role of the 2 pairs of fibrinopeptides. *Biophys. J.* 1986;50:1079-93.

[62] Weisel JW, Veklich Y, Gorkun O. The sequence of cleavage of fibrinopeptides from fibrinogen is important for protofibril formation and enhancement of lateral aggregation in fibrin clots. *J. Mol. Biol.* 1993;232:285-97.

[63] Blomback B. Fibrinogen and fibrin-proteins with complex roles in hemostasis and thrombosis. 1996, 83, 1-75.

[64] Blomback B, Hessel B, Hogg D, Therkildsen L. 2-step fibrinogen-fibrin transition in blood coagulation. *Nature.* 1978, 275, 501-505.

[65] Shen LL, Hermans J, McDonagh J, McDonagh RP. Role of fibrinopeptide B release: comparison of fibrins produced by thrombin and Ancrod. *Am. J. Physiol.* 1977;232:H629–33.

[66] Torbet J. Fibrin assembly after fibrinopeptide A release in model systems and human plasma studied with magnetic birefringence. *Biochem. J.* 1987;244:633–7.

[67] Blomback B, Carlsson K, Fatah K, Hessel B, Procyk R. Fibrin in human plasma: gel architectures governed by rate and nature of fibrinogen activation. *Thromb. Res.* 1994;75:521–38.

[68] Fredenburgh JC, Stafford AR, Pospisil CH, Weitz JI. Modes and consequences of thrombin's interaction with fibrin. Biophys. Chem. 2004;112:277-84.

[69] Di Cera E. Thrombin interaction. *Chest*. 2003;124:11S-17S.

[70] 28 Binnie CG, Lord ST. The fibrinogen sequences that interact with thrombin. *Blood*.1993;81:3186-92.

[71] Pechik I, Madrazo J, Mosesson MW, Hernandez I, Gilliland GL, Medved L. Crystal structure of the complex between thrombin and the central "E" region of fibrin. *P. Natl. Acad. Sci. U.S.A.* 2004;101:2718-23.

[72] Mullin JL, Gorkun OV, Binnie CG, Lord ST. Recombinant fibrinogen studies reveal that thrombin specificity dictates order of fibrinopeptide release.
J. Biol. Chem. 2000;275:25239-46.

[73] Pechik I, Yakovlev S, Mosesson MW, Gilliland GL, Medved L. Structural basis for sequential cleavage of fibrinopeptides upon fibrin assembly. *Biochemistry*. 2006;45:3588-97.

[74] Kita R, Takahashi A, Kaibara M, Kubota K. Formation of fibrin gel in fibrinogen-thrombin system: static and dynamic light scattering study. *Biomacromolecules*. 2002;3(5):1013-1020.

[75] Rocco M, Bernocco S, Turci M, Profumo A, Cuniberti C, Ferri F. Early events in the polymerization of fibrin. *Ann. N.Y. Acad. Sci.* 2001;936:167- 184.

[76] Chernysh IN., Nagaswami C.,. Weisel JW, Visualization and identification of the structures formed during early stages of fibrin polymerization, *Blood*. 2011 117: 4609-4614.

[77] Weisel JW, Veklich Y, Gorkun O. The sequence of cleavage of fibrinopeptides from fibrinogen is important for protofibril formation and enhancement of lateral aggregation in fibrin clots. *J. Mol. Biol.* 1993;232, 285-297.

[78] Hantgan R, Fowler W, Erickson H, Hermans J. Fibrin assembly: a comparison of electron microscopic and light scattering results. *Thromb Haemost*. 1980;44(3):119-124.

[79] Doolittle RF. Fibrinogen and fibrin. *Annu. Rev. Biochem.* 1984;53:195-229.

[80] Chernysh IN., Nagaswami C., Weisel JW., Visualization and identification of the structures formed during early stages of fibrin polymerization, *Blood*. 2011, 117: 4609-4614.

[81] Blomback B, Carlsson K, Hessel B, Liljeborg A, Procyk R, Aslund N. Native fibrin gel networks observed by 3D microscopy, permeation and turbidity. *Biochim. Biophys. Acta.* 1989;997(1-2):96- 110.

[82] Chernysh IN, Weisel JW. Dynamic imaging of fibrin network formation correlated with other measures of polymerization. *Blood.* 2008;111(10): 4854-4861.

[83] Janmey PA, Ferry JD. Gel formation by fibrin oligomers without addition of monomers. *Biopolymers.* 1986;25(7):1337-1344.

[84] Carr ME, Shen LL, Hermans J. Mass-length ratio of fibrin fibers from gel permeation and light scattering. *Biopolymers.* 1977;16:1–15.

[85] Wolberg AS, Monroe DM, Roberts HR, Hoffman M. Elevated prothrombin results in clots with an altered fiber structure: a possible mechanism of the increased thrombotic risk. *Blood.* 2003;101:3008–13.

[86] Hartmann A, Boukamp P, Friedl P. Confocal reflection imaging of 3D fibrin polymers. *Blood Cells Mol. Dis.* 2006;36:191–3.

[87] Nemerson Y, Turitto VT. The effect of flow on hemostasis and thrombosis. *Thromb. Haemost.* 1991; 66: 272–6.

[88] Hathcock JJ, Nemerson Y. Platelet deposition inhibits tissue factor activity: in vitro clots are impermeable to factor Xa. *Blood.* 2004; 104: 123–7.

[89] Mosesson MW. Antithrombin I. Inhibition of thrombin generation in plasma by fibrin formation. *Thromb. Haemost.* 2003; 89: 9–12.

[90] Stubbs MT,Bode W. A player of many parts: The spotlight falls on thrombin's structure. *Thromb. Res.* 1993; 69; 1-58.

[91] Weisel JW. The electron microscope band pattern of human fibrin: Various stains, lateral order, and carbohydrate localization. *J. Ultrastruct. Mol. Struct. Res.* 1986; 96; 176-188.

[92] Di Cera E. Thrombin interactions. *Chest.* 2003; 124; 11-17.

[93] Rubens FD,Weitz JI,Brash JL,Kinlough-Rathbone RI. The effect of antithrombin III-independent thrombin inhibitors and heparin on fibrin acceretion onto fibrin-coated polyethylene. *Thrombosis Haemostasis.* 1993; 69: 130-134.

[94] Riedel T., Brynda E., Dyr JE., Houska M, Controlled preparation of thin fibrin films immobilized at solid surfaces, *J. Biomed. Mater. Res.*, 88A, 2009, 437-447.

[95] Brynda E.,Tsepalova NA., Štol M., Equilibrium adsorption of human serum albumin and human fibrinogen on hydrophobic and hydrophilic surfaces, *J. Biomed. Mater. Res.* 1984, 18, 685-93.

[96] Rodriguez Emmenegger C., Brynda E., Riedel T., Sedlakova Z., Houska M., Bologna A., Interaction of blood plasma with anti-fouling surfaces., *Langmuir.* 2009, 25(11),6328–6333.

[97] M.. Lestelius, B. Liedberg, P. Tengvall, In Vitro Plasma Protein Adsorption on ö-Functionalized Alkanethiolate Self-Assembled Monolayers, *Langmuir*. 1997, 13, 5900–5908.

[98] Tang L., Eaton JW., Fibrin(ogen) mediates acute inflammatory responses to biomaterials. *J. Exp. Med.*, 178, 2147–2156 (1993).

[99] Laurens N, Koowijk P, De Maat MPM. Fibrin structure and wound healing. *Journal of Thrombosis and Haemostasis*. 2006, 4, 932-939.

[100] Geer CB, Tripathy A, Schoenfisch MH, Lord ST, Gorkun OV. Role of B-b knob-hole interactions in fibrin binding to adsorbed fibrinogen. *J. Thromb. Haemost.* 2007;5(12):2344-2351.

[101] Geer CB, Rus IA, Lord ST, Schoenfisch MH. Surface-dependent fibrinopeptide A accessibility to thrombin. *Acta Biomater*. 2007;3(5):663-668.

[102] Evans-Nguyen KM, Fuierer RR, Fitchett BD, Tolles LR, Conboy JC, Schoenfisch MH. Changes in adsorbed fibrinogen upon conversion to fibrin. *Langmuir*. 2006;22(11):5115-5121.

[103] Soman P, Rice Z, Siedliecki CA. Measuring the time-dependent functional activity of adsorbed fibrinogen by atomic force microscopy. *Langmuir*. 2008;24(16):8801-8806.

[104] Savage B, Ruggeri ZM. Selective recognition of adhesive sites in surface-bound fibrinogen by glycoprotein IIb-IIIa on nonactivated platelets. *J. Biol. Chem.* 1991;266(17):11227-11233.

[105] Hu WJ, Eaton JW, Ugarova TP, Tang L. Molecular basis of biomaterial-mediated foreign body reaction. *Blood*. 2001;98(4):1231-1238.

[106] Dyr JE, Tichý I, Jiroušková M, et al. Molecular arrangement of adsorbed fibrinogen molecules characterized by specific monoclonal antibodies and a surface plasmon resonance sensor. *Sensor Actuat. B-Chem*. 1998;51(1-3):268-272.

[107] Dyr JE, Cajthamlova H, Suttnar J, Fortova H. Fibrinopeptide release from surface-bound fibrinogen. In: Ernst E., Koenig W, Low GDO, Meade TW, eds. Fibrinogen : A "new" cardiovascular risk factor. Wien, Blackwell-MZV, 1992: 27-30.

[108] Brynda E, Houska M, Lednicky F. Adsorption of human fibrinogen onto hydrophobic surfaces: The effect of concentration in solution. *J. Colloid Interf. Sci.* 1986;113(1):164-171.

[109] Riedel T, Suttnar J, Brynda E, Houska M, Medved L, Dyr JE. Fibrinopeptides A and B release in the process of surface fibrin formation. *Blood*. 2011, 117 (11) 6328-6333.

[110] Mullin JL, Gorkun OV, Binnie CG, Lord ST. Recombinant fibrinogen studies reveal that thrombin specificity dictates order of fibrinopeptide release. *J. Biol. Chem.* 2000;275(33):25239-25246.

[111] Dyr JE, Blombäck B, Hessel B, Kornalík F. Conversion of fibrinogen to fibrin induced by preferential release of fibrinopeptide B. *Biochim. Biophys. Acta.* 1989; 990, 18-24.

[112] Blombäck B, Bark N. Fibrinopeptides and fibrin gel structure. *Biophys. Chem.* 2004;112(2-3):147-151.

[113] Blomback B. Hessel B., Okada M., Egberg N, Mechanism of fibrin formation and its regulation, *Ann. N.Y. Acad. Sci.* 370 (1981) 536– 544.

[114] Sporn LA, Bunce LA, Francis CW. Cell proliferation on fibrin: modulation by fibrinopeptide cleavage. *Blood.* 1995;86(5):1802-1810.

[115] Filová E, Brynda E., Riedel T., Bačáková L., Chlupáč J., Lisá V., Houska M., Dyr JE., Vascular Endothelial Cells on Two- and Three-Dimensional Fibrin Assemblies for Biomaterial Coatings, *J. Biomed. Mater. Res.*, 90 (2009) 55-69.

INDEX

C

N

O

P

Q

R

S

T

U

V

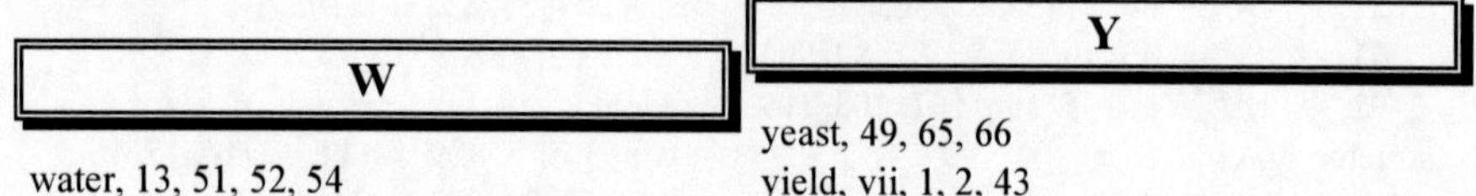
W

Y